Immunochemistry of AIDS

Chemical Immunology

Vol. 56

Basel · Freiburg · Paris · London · New York · New Delhi · Bangkok · Singapore · Tokyo · Sydney

Immunochemistry of AIDS

Volume Editor
Erling Norrby, Stockholm

18 figures, 6 color plates and 12 tables, 1993

Basel · Freiburg · Paris · London · New York · New Delhi · Bangkok · Singapore · Tokyo · Sydney

Chemical Immunology

Formerly published as 'Progress in Allergy'
Founded 1939 by Paul Kallòs

Bibliographic Indices

This publication is listed in bibliographic services, including Current Contents® and Index Medicus.

Drug Dosage

The authors and the publisher have exerted every effort to ensure that drug selection and dosage set forth in this text are in accord with current recommendations and practice at the time of publication. However, in view of ongoing research, changes in government regulations, and the constant flow of information relating to drug therapy and drug reactions, the reader is urged to check the package insert for each drug for any change in indications and dosage and for added warnings and precautions. This is particularly important when the recommended agent is a new and/or infrequently employed drug.

Printed on acid-free paper
ISBN 3–8055–5655–1

Contents

Introduction IX

Genotypic and Phenotypic Variation of HIV-1: Impact on AIDS Pathogenesis and Vaccination
Wolfs, T.F.W. (Amsterdam); Nara, P.L. (Frederick, Md.); Goudsmit, J. (Amsterdam) 1

Origin of Retroviruses 1
Phylogeny of the Lentiviral Group 3
Genetic Variability of HIV-1 5
Mechanism of Variation 5
Variation among HIV Isolates and within an Isolate 6
Rate of Fixation of Mutations 8
Phenotypic Variability of HIV-1 9
Biological Variation 9
Antigenic Variation 9
Role of the V3 Domain in HIV Infection 10
Structure of the V3 Domain 10
Principal Neutralization Domain 10
Determinant of Biological Phenotype 12
Impact of Variation of the Third Variable gp120 Region during Natural HIV-1 Infection 13
The Acute Phase of HIV-1 Infection 13
The Asymptomatic Phase 14
The Symptomatic Phase 16
Implication of Viral Diversity for Vaccine Development 18
References 20

B Cell Antigenic Site Mapping of HIV-1 Glycoproteins
Neurath, A.R. (New York, N.Y.) 34

Amino Acid Sequence Variability of HIV-1 gp120/gp41 38
Mapping of B Cell Epitopes on gp120/gp41 Using Peptides 38

gp120/gp41 B Cell Epitope Mapping Using Antipeptide Antisera 41
Synthetic Peptides as Diagnostic Reagents to Detect and Subtype Anti-HIV-1 Antibodies 42
Site-Directed Serology for Detection of HIV Infections Using Synthetic Peptides Encompassing the 605–611 Disulfide Loop of gp41 42
Synthetic Peptides as Site-Specific Reagents for Subtyping of Anti-HIV-1 Antibodies 44
Synthetic Peptides as Potential Components of Anti-HIV-1 Vaccines 45
Discontinuous Virus-Neutralizing Epitopes 47
Molecular Mimicry of gp120/gp41 B Cell Epitopes 49
Conclusions 49
References 50

Antigenic and Immunogenic Sites of HIV-2 Glycoproteins
Chiodi, F.; Björling, E.; Samuelsson, A.; Norrby, E. (Stockholm) 61

Epidemiological Features and Natural History of HIV-2 Infection 62
Genomic Organization and Immunological Relatedness of HIV-1 and HIV-2 . . 62
Animal Models for HIV Vaccine Studies 63
Antibody-Binding Antigenic Sites in the HIV-2 Glycoproteins 65
Strain- and Type-Cross-Reactive Neutralization of HIV-2 Antibody Positive Sera 67
Target Regions for Neutralizing Antibodies in the HIV-2 Glycoproteins 68
Concluding Remarks 71
References 72

SIV Neutralization Epitopes
Javaherian, K. (Cambridge, Mass.); Langlois, A.J. (Durham, N.C.); LaRosa, G.J. (Cambridge, Mass.) 78

Some Structural Characteristics of the SIV Envelope 79
Neutralization Domain of the SIV Envelope 83
Conclusion 88
References 88

Epitopes of HIV-1 Glycoproteins Recognized by the Human Immune System
Laal, S.; Zolla-Pazner, S. (New York, N.Y.) 91

Epitopes on gp120 91
Epitopes on gp41 100
Concluding Remarks 105
Acknowledgements 106
References 106

Human Antibodies to HIV-1 by Recombinant DNA Methods
Burton, D.A.; Barbas, C.F., III (La Jolla, Calif.) 112

Phage Display of Antibody Combinatorial Libraries and Selection of Specific Fab Fragments 114
Application of the Combinatorial Library Approach to the Generation of Antibodies to HIV-1 116
Neutralization of HIV-1 by Recombinant Fabs 119
Genetic Manipulation of Antibodies 120
Gene Rescue from Antibody-Producing Cell Lines 122
Alternatives to Seropositive Humans as Sources of Antibody Libraries 122
Acknowledgements 124
References 124

CD4+ T Cell Epitopes in HIV-1 Proteins
Siliciano, R.F. (Baltimore, Md.) 127

General Aspects of the CD4+ T Cell Response to HIV-1 127
The Importance of Defining CD4+ T Cell Epitopes 128
The CD4+ T Cell Response to HIV-1 in Natural Infection 129
Role of the Presenting MHC Molecule 131
Approaches to Identifying CD4+ T Cell Epitopes 133
Methods for Predicting T Cell Epitopes 134
Clonal Analysis of Human CD4+ T Cell Epitopes 135
Analysis of Peptide-Specific T Cell Responses in Infected Individuals 140
Delineation of T Cell Epitopes in Animal Models 143
Delineation of Epitopes by Sequence Analysis of Peptides Eluted from MHC Molecules 143
Conclusions 144
References 145

How Does the HIV Escape Cytotoxic T Cell Immunity?
Phillips, R.E.; McMichael, A.J. (Oxford) 150

Immunity to Viruses 151
Cytotoxicity T Cell Recognition of Viruses 153
Antigen Processing and HLA Class I Binding 154
Cytotoxic T Cells Specific for HIV 155
Genetic Variation in the Cytotoxic Epitopes of HIV gag 157
Conclusion 159
References 160

Subject Index 165

Introduction

Research aiming at understanding and managing infectious diseases is unique in many aspects. Its focussing ranges from studies of molecules to characterization of epidemics. From another perspective, a separate but still intertwined analysis needs to be done of each of the two combatants, the parasite and the host. The infection with human immunodeficiency virus (HIV) represents a formidable medical problem and it has a spectrum of qualities, promptly turning our attempts to understand the pathogenic process into a major intellectual challenge. One of the central issues in understanding the disease is its immunology. Why is it that the immune system can effectively suppress the primary infection, but that eventually the infection comes back and leads to a fatal outcome? Which are the key problems in the unsuccessful attempts of the immune system to clear the infection and in the progressive erosion of the immune defence during the eclipse period? Finally, from a practical vantage point, how can the immune defence be exploited to either prophylactically manage the primary infection or therapeutically guard against the resurge of the endogenous infection?

To state that a book on the immunochemistry of acquired immunodeficiency syndrome (AIDS) is timely, is a truism. The rapid accumulation of new knowledge continuously opens new perspectives. In this process it is of interest to attempt deducing at which stage of development time is ripe for turning knowledge into practical application. It is possible that we are now at a critical stage. A remarkably penetrating knowledge of both B cell and T cell functions in HIV immunology and immunochemistry has accumulated as evidenced by the reviews in this book. Attempts are being made to design synthetic vaccines. Furthermore, a dramatic breakthrough has been made in exploiting the recombinant DNA methods to select human antibody structures of defined specificities at will. In the near future, the potential therapeutic use of such reagents in both lentivirus-infected animals and in HIV-infected human beings will be evaluated.

It is my hope that the readers will enjoy these state-of-the-art presentations by the foremost specialists in the field and that thoughts generated by reading the different contributions will spur further research that can help to bring the AIDS epidemic under control.

Erling Norrby

Norrby E (ed): Immunochemistry of AIDS.
Chem Immunol. Basel, Karger, 1993, vol 56, pp 1–33

Genotypic and Phenotypic Variation of HIV-1: Impact on AIDS Pathogenesis and Vaccination

Tom F.W. Wolfs[a], *Peter L. Nara*[b], *Jaap Goudsmit*[a]

[a] Department of Virology, University of Amsterdam, Academic Medical Centre, Amsterdam, The Netherlands; [b] Virus Biology Unit, LTCB, BCP, DCE, NCI-FCRDC, NIH, Frederick, Md., USA

Knowledge of the evolution and variation of the human immunodeficiency virus (HIV) is a prerequisite for understanding the pathogenesis and design of intervention against HIV infection. This review gives a short introduction to the origin of HIV and its relationship with other members of the retroviruses. The genetic variation of HIV and its impact on phenotypical properties are described. Subsequently, the third variable envelope region is introduced as the principal neutralization domain and an important determinant of biological phenotype. Finally, the impact of HIV-1 variation on the natural history of acquired immunodeficiency syndrome (AIDS) and on preventive strategies is described.

Origin of Retroviruses

RNA viruses have been designated retroviruses when they exhibit the ability to copy their RNA genome into double-stranded DNA that can be integrated into the DNA of an infected cell [1, 2]. This conversion from RNA to DNA is catalyzed by an RNA-directed DNA polymerase or reverse transcriptase (RT). Retroviruses, however, are not unique in having RT-coding sequences. Many other RT-bearing entities are described, including mobile elements found in a wide variety of eukaryotes, some plant and animal DNA viruses and even certain fungal group II mitochondrial introns and a mitochondrial plasmid [3, 4]. To date, there is no universal agreement about the origin and evolution of these RT-containing genetic elements, which are referred to as retroelements. The protovirus hypothesis, first put forward by Temin in 1970 [5], suggests that retroviruses evolved from cellular transposable elements. In 1989, bacterial RT (the so-called retron) [6, 7] was discovered, suggesting that enzymes with reverse transcription activity are more ancient than the separation between prokaryotes and eukaryotes. The retron, being the shortest and simplest of the retroelements, was implicated as the precursor of all present retroelements [8].

An alternative model, based on comparison of the genetic organization of RT-containing elements and RNA-directed RNA polymerases from various plus-strand RNA viruses, suggests a common ancestry of all current retroelements from a transposable element with both *gag*-like and *pol*-like genes [9]. In this way, retroviruses represent transposable elements which have acquired an envelope *(env)* gene, allowing them to leave the cell and subsequently infect new cells. Following this alternative hypothesis, retroelements of organelle and bacterial genomes are not the precursor of retroviruses but another descendent from the common ancestor, resulting from a transfer of the RT domain of the *pol* gene from the ancestor to a mitochondrial plasmid, the mitochondrial genome or the bacterial genome.

Retroviruses exist in a multitude of forms involving varying degrees of infectivity and pathogenicity. Endogenous retroviruses are distinguished from exogenous retroviruses in that they are usually carried benignly in the germ line. Most of these endogenous retroviruses are defective [10].

According to RT sequence similarity, exogenous retroviruses fall into four major groups: the murine leukemia virus group; the mouse mammary tumor virus/Rous sarcoma virus group; the human T cell leukemia virus (HTLV) group, and the lentivirus group [11]. Extensive homologies exist between these exogenous retroviruses and endogenous retroviruses; human and baboon endogenous retroviruses (HERV-E, HERV-K, BaEV) are clustered in the first two groups. There are two reasons for not expecting to find this degree of homology. One, because the present day strains of exogenous retroviruses are introduced only recently, whereas endogenous sequences are deeply rooted in primate evolution. One endogenous retrovirus (HERV-C) has copies distributed throughout the genome in exactly the same places in both humans and chimpanzees [12], indicating that this retrovirus was introduced into the genome before chimpanzees and humans diverged more than 8 million years ago. Others (including HERV-E) have been present in the primate germ line for at least 30 million years [13]. Two, because the mutation rate of exogenously replicating retroviruses is at least 10^5 times greater than that of endogenously replicating retroviruses (replicating at the highest cellular gene or pseudogene rate). To explain this discrepancy,

Doolittle et al. [10] proposed a model whereby 'exogenous retroviruses exist as short-lived bursts upon a backdrop of germline-encoded endogenous retroviruses'. In other words, the pool of endogenous retroviral sequences periodically contributes to the generation of exogenous viruses. In humans, the finding of HIV-1-, HTLV-I- and HTLV-II-like endogenous sequences adds to the credibility of this model [14, 15].

Phylogeny of the Lentiviral Group

Three recent outbreaks of primate lentiviruses have been recognized: HIV-1 in Central Africa, Asia, North America and Europe; HIV-2 in West Africa, and simian immunodeficiency virus (SIV) in North American primate colonies.

The first cases of AIDS were described among homosexual men in the United States in 1981 [16–19]. Following these reports, AIDS cases were described among persons with hemophilia [20], blood transfusion recipients [21], heterosexual intravenous drug users [22], infants born to infected mothers [23] and partners of infected individuals [24]. The causative agent of AIDS was identified in 1983, when Montagnier and coworkers isolated a retrovirus [25]. The virus, originally called lymphadenopathy-associated virus (LAV), is now designated HIV-1 [26]. Although the first cases of AIDS were reported in the United States, the epidemic is now worldwide, with the highest prevalence in endemic areas in Africa. The highest incidence is currently in Brazil, Central Africa and Asia [27].

In 1986, Clavel et al. [28] isolated a retrovirus related to HIV-1 from 2 West African patients with AIDS. Antibodies raised against HIV-1 could immunoprecipitate the *gag* and *pol* products of these isolates but not the *env* products, whereas previous HIV isolates induced antibodies cross-reactive with divergent HIV-1 envelope glycoproteins. Moreover, in conditions of very low stringency the genome of these new viruses cross-hybridized only poorly with HIV-1 DNA probes [28, 29]. The new virus originally called LAV-2 is therefore now referred to as HIV-2.

In the early 1980s, a cluster of lymphomas and immunodeficiency-associated disorders (similar to AIDS in humans) was identified in a colony of captive macaques in a North American primate center [30]. SIV-mac was isolated from these macaques and identified as the causative agent [31, 32]. In the wild, however, macaques do not appear to be infected with SIV [33, 34].

After determination of the first sequences of HIV-1 [35], HIV-2 [36] and SIV-mac isolates [37, 38], comparative analysis showed that HIV-1 differed significantly in nucleotide sequence from HIV-2: overall, they displayed only 42% homology [36]. In contrast, HIV-2 was closely related to SIV-mac [37–39]. The observation that healthy sooty mangabeys – monkeys that exclusively inhabit West Africa (the location of the HIV-2 epidemic) – were infected with SIV (SIV-sm) both in captivity in North American primate centers [40, 41] and in the wild [42], and the fact that SIV-sm proved to be closely related to SIV-mac and HIV-2 [42], provided a possible explanation for the close relationship between the endemic HIV-2 in West Africa and the SIV-mac found in the captive macaques in North America. HIV-2 and SIV-mac may have resulted from cross-species transmission of SIV-sm from sooty mangabeys to humans and macaques, respectively. West African sooty mangabeys were imported into the United States up to 1968 and may have been housed together with macaques [42]. After the description of the isolation of an HIV-1-like lentivirus from chimpanzees (SIV-cpz) [43, 44], a similar hypothesis for the introduction of HIV-1 in Central Africa has been put forward. SIV-cpz may have evolved as HIV-1 after successful infection of a human. Of all the SIVs so far reported, SIV isolated from African green monkeys (SIV-agm) displayed the highest genetic diversity [45]. African green monkeys were already widely infected in the 1950s [46] but do not develop the AIDS-like illness. Taken together, the high genetic variability and the early documentation of infection suggest that SIV-agm may be the oldest primate lentivirus in existence [45, 47]. Another SIV was isolated in Africa from mandrils (SIV-mnd) [48]. SIV-mnd was shown to be a distinct, nonhuman primate lentivirus. Thus, among these primate lentiviruses, four main groups can be distinguished: HIV-1/SIV-cpz; HIV-2/SIV-mac/SIV-sm; SIV-agm, and SIV-mnd.

It has been estimated that all known primate lentiviruses could have evolved from a common ancestor that existed 140 [49] to 280 [50] years ago. This estimate is based on the phylogenetic analysis of viral proteins and the assumption that genes evolved at a certain constant rate. The separation from visna, a nonprimate lentivirus causing degenerative disease of the central nervous system in sheep, is estimated to have occurred approximately 300 years ago [49]. The divergence of SIV and HIV-2 must have occurred approximately 30 years ago [42, 49, 51, 52], at the same time as the SIV-cpz HIV-1 divergence [52]. Since the most divergent HIV-1 isolate, Z3, diverged from the other HIV-1 isolates around 1960 [53], the ancestral HIV-1 virus must have existed before 1960. This is consistent with epidemiological data:

the earliest positive HIV-1 response was found in serum collected in Zaire in 1959 [54]. Eigen and Nieselt-Struwe [55] used a new method to calculate evolution times from sequence distance. This method not only specified the three codon positions and distinguished transitions from transversions, but also assigned individual probabilities of substitution (constant, variable and hypervariable) to all nucleotide positions. The result of this analysis drastically lengthened divergence times. Calculated in this way, the oldest node linking all human and simian viral sequences dates back at least 600–1,200 years.

Genetic Variability of HIV-1

Mechanism of Variation

The extent of lentivirus variation ('the mutation rate') depends on the number of replication cycles, the growth rate of the viral population and the infidelity of the viral polymerases. The rate of fixation of these mutations in the population ('the evolution rate') is dependent not only on this mutation rate, but also on the positive selection for variation of environmental conditions (such as immune pressure and antiviral drugs) and the negative selection against variation imposed by the functional constraints of the virus.

The molecular basis for the variation is the viral polymerase error occurring in the absence of exonucleotic activity (proofreading). Replication of HIV involves three steps: reverse transcription; plus-strand DNA synthesis, and transcription. Higher HIV-1 RT infidelity compared to other RTs has been reported by some researchers [56–58] but not by others [59]. Error frequency rates originally determined on DNA templates range from 1/1,700 to 1/7,400 nucleotides [56, 57, 60] and do not differ significantly from RNA templates [58]. Translated to one replication cycle, this would mean 2.5–10 errors per genome for HIV. The rate of misincorporation is not uniform across the whole genome but seems to be highly dependent on the specific context. A:G mismatches, for example, occur frequently when the template adenine residue is preceded by the doublets AT, TT, CT or AC and are prevented when the preceding doublets are NG, GT, TC or GC [59]. In addition to single-base substitutions, minus-one frameshifts and sequence changes involving more than one nucleotide have been reported to occur frequently during DNA synthesis by RT [61].

Recombination adds further to the complexity of genetic variation by spreading variation from one genome to another. Retrovirus recombination, which requires a heterozygous virion [62], can be highly efficient. Based on observed replication rates in an experimental system, Hu and Temin [62] estimated that one of three 10-kb viruses would experience at least one recombination event [62]. Two models for retrovirus recombination have been put forward. Coffin [63] suggested a switch of template during minus-strand generation. Minus-strand recombination may increase the virus variability by producing a normal recombinant from two RNA molecules with different deleterious mutations. Junghans et al. [64] proposed a crossing-over of the DNA plus-strand onto two DNA minus-strand copies. The latter mechanism would result in a heterozygous provirus. However, such a heteroduplex is in fact always repaired before integration [65].

Variation among HIV Isolates and within an Isolate

Shaw et al. [66, 67] and Hahn et al. [68, 69] were the first to describe the genomic diversity of HIV-1 (then called HTLV-III) viruses. Using restriction enzyme mapping and heteroduplex thermal melt analysis, they showed that HIV isolates from patients with AIDS or AIDS-related complex (among which MN and RF) and molecularly cloned genomes from one patient isolate (IIIB) demonstrated variation, which was located largely in the envelope coding region [66–69]. In a more extensive study, each of 18 isolates from individuals with AIDS showed a different restriction enzyme pattern [70]. Most of the differences were found in the envelope coding region. The complete nucleotide sequences of isolate IIIB [35], LAV [71] and ARV-2 [72] gave further evidence for the heterogeneity of HIV-1 viruses. Sequenced genomes of two isolates obtained from Zairian patients (ELI and MAL) revealed a much greater extent of genetic polymorphism than previously observed [73], although the genetic organization was identical to that of isolates from Europe and North America. Sequences of the African isolates were more divergent from the European/North American isolates than any of the isolates described thus far. In addition, the divergence between the two African isolates ELI/MAL was similar to the divergence between ELI/LAV and MAL/LAV, suggesting a longer evolution of the virus in Africa.

All these studies found that the *env* gene was more variable than the *gag* and *pol* genes. A distinct pattern of variation was evident within the *env* gene. Based on comparison of several sequences, five hypervariable regions (V1–V5) were defined as regions exhibiting less than 30% amino acid conservation [74, 75]. These regions were separated by regions with a higher degree of conservation. The variable regions coincided with the predicted antigenic sequences [76].

Hahn et al. [77] reported genetic variation over time in viruses isolated from three individuals. Changes in the *env* genes were almost exclusively point mutations. All these changes were clustered in the hypervariable regions of the envelope previously described. Variation between viral isolates from one individual was always less than the variation between isolates from different individuals. Extended studies by the same group showed that internal variation within HIV isolates was extensive and that viral isolates consisted of a complex mixture of genotypically distinguishable viruses, unless biologically or molecularly cloned [78]. Fisher et al. [79] confirmed these observations. Chimeric viruses were generated in which the envelope region of the original (HXB2) virus was substituted by that of six viral clones from one viral isolate. The resulting viruses differed widely in their capacity to grow in human T cells, cell lines and monocytoid cultures. Apparently, the biological properties of molecularly cloned viruses from one isolate differed substantially. Using the polymerase chain reaction [80], the real extent of variation became apparent. Every genome has to be considered unique and consequently HIV isolates must be described in terms of populations of closely related genomes or as a quasispecies [81].

The quasispecies concept, based on Darwin's theory of natural selection, was introduced by Eigen et al. in 1971 [82, 83]. Molecules that replicate with a limited fidelity generate enough diversity in genotypes to contain some that code for adaptively beneficial phenotypic changes. Natural selection results in the survival of the fittest mutant (or the most efficient at producing offspring). Because of the highly error-prone replication machinery, not only this mutant but a whole swarm of genotypes containing a large number of mutants in varying frequencies survives. The representation of each mutant in such a swarm is determined by the rate at which each mutant arises and its relative fitness. Eigen named this mutant distribution (often characterized by the most frequent or 'master' sequence) a quasispecies distribution. HIV quasispecies distribution, stabilized around the master sequence but containing a swarm of mutants, enables the virus to adapt to the host rapidly. If the environmental conditions are such that a previously unfavored mutant is now selected, a shift in quasispecies distribution will take place. The new distribution will be centered around this newly selected, 'fittest' mutant. The establishment of such a new quasispecies is what we call evolution. For HIV, the complexity of the quasispecies showed no correlation with the stage of disease [81].

This in vivo diversity is presumably lost in vitro by the selective outgrowth of HIV-1 strains capable of rapid replication. This phenomenon was first described by Meyerhans et al. [84], who studied sequence diversity of the *tat* gene from an HIV-1-infected homosexual man over a 2.5-year period, before and after culture amplification. Fluctuations in the in vivo quasispecies were not reflected in the quasispecies of the in vitro samples. Moreover, sequence populations were more complex in vivo than in vitro. This effect of culturing repeatedly observed also by others [85–89] warranted the conclusion that 'to culture is to disturb' [84]. Sequence variation of HIV isolates between patients infected from a common source has been reported [90–94]. Although interpatient variability was extensive [90–92], it was markedly less extensive than the variability found in epidemiologically unlinked individuals. The viral heterogeneity within such a cohort seemed to increase over time [92].

Within an individual, differences in the frequency and distribution of sequence variants were noted in brain tissue versus blood cells [87, 95]. This may indicate a separate, tissue-specific evolution of the viral quasispecies. In addition, within the peripheral-blood compartment, differences were demonstrated between the viral RNA from cell-free virus particles circulating in plasma and proviral DNA present in HIV-infected peripheral-blood mononuclear cells (PBMCs). Sequence changes were found in both, but novel variants initially appeared in the plasma RNA populations and subsequently became detectable in the PBMCs [96].

Rate of Fixation of Mutations

As discussed previously, the evolution rate observed in a viral population is dependent on the rate of mutation and the selection force on the mutated population. Hahn et al. [77] estimated the nucleotide substitution rate of the *gag* gene to be between 0.4×10^{-3}/site/year and 1.9×10^{-3}/site/year and that of the *env* gene between 3.2×10^{-3}/site/year and 15.8×10^{-3}/site/year. These estimates were based on sequence comparisons of sequentially isolated viruses. Importantly, they also showed that evolutionary change was not a linear process. Although viruses were obtained sequentially, the viruses had not evolved genetically in that order. Rather, isolates seemed to have evolved in parallel. In a hemophiliac cohort, the rate of sequence change in three variable envelope regions averaged 4.5×10^{-3} nucleotide/site/year [90]. Li et al. [53] used complete sequences of published isolates to calculate evolution rates. The highest rates were observed for the hypervariable envelope regions (14.0×10^{-3} nucleotide change/site/year for nonsilent sites and 17.2×10^{-3} nucleotide change/site/year for silent sites). The *pol* region was the most conservative site of the viral genome (1.7×10^{-3} nucleotide change/site/year).

Due to the enormous diversity of viral genomes as well as their ability to change rapidly, quasispecies distribution enables the virus to undergo fast biological and antigenic variation.

Phenotypic Variability of HIV-1

Biological Variation

HIV-1 variants have been shown to display variation in their biological properties. Isolates can be distinguished on the basis of their replication rate, their capacity to infect different cell types and their capacity to induce syncytia [97–102]. Isolates recovered from individuals with AIDS or AIDS-related complex showed higher replication rates than isolates recovered from asymptomatic individuals. In contrast to viruses isolated from asymptomatic individuals, they were able to establish persistent infection in T cell lines [98]. The ability to form syncytia in primary PBMC cultures was a third biological property found more frequently the more advanced the disease [103, 104]. In longitudinal studies, the transition from non-syncytium-inducing (non-SI) to syncytium-inducing (SI) viral isolates correlated with disease progression [105–107]. A strict correlation was found between the ability to induce syncytia in PBMC cultures and the ability to infect T cell lines continuously and productively [106, 107]. Non-SI isolates, however, appeared to be much more monocytotropic than SI isolates and were not transmissible to T cell lines [108–109]. The biological phenotypes described are those of uncloned primary isolates. Primary isolates showing the non-SI phenotype exclusively contain clones with the non-SI phenotype. In contrast, SI isolates nearly always constitute a mixed population of clones with and without SI capacity [110–112]. Apparently, the phenotype of the primary isolate is determined by the phenotype of the most virulent clone present.

It is argued that non-SI monocytotropic clones are responsible for the persistence of HIV infection. Progression of disease is associated with a selective increase of T-cell-tropic, non-monocytotropic clones [112].

Antigenic Variation

HIV isolates with a distinct sensitivity to serum neutralization have been reported [113]. In vitro, neutralizing antibodies generated neutralization-resistant viruses [114, 115]. In an experimental HIV-1 infection of two chimpanzees, neutralization-resistant viruses emerged as early as 16 weeks after inoculation, whereas in vitro culture resulted in an outgrowth of the most replication-competent strain of the inoculum, but not in neutralization resistance [116]. A similar observation was made in a study by Albert et al. [117] in which four patients were followed from the moment of acute HIV-1 infection on. Viruses isolated early in infection were neutralized by both early and late sera. However, virus variants isolated at later stages (more than 6 months after the primary infection) were neutralized poorly or not at all. Antigenic divergence has been described extensively in other lentiviral infections, such as visna [118–120] and equine infectious anemia virus [121, 122].

Role of the V3 Domain in HIV Infection

Structure of the V3 Domain

The V3 domain generally consists of about 35 amino acids flanked by two cysteines that form a disulfide bridge [123]. Although the region is variable, several conserved sequence and structural elements were found in a large set of V3 sequences from different isolates [124]. Based on a neural network approach, it has been predicted that the average V3 loop structure is: cysteine – β-strand – type II β-turn – β-strand – α-helix – cysteine [124]. The reverse β-turn is formed by the highly conserved triplet Gly-Pro-Gly on top of the loop. Nuclear magnetic resonance analysis of free V3 peptides in aqueous solution has produced no evidence for the ordered β-strands and α-helices, but the predicted reverse β-turn at the top of the loop was observed [125].

Principal Neutralization Domain

Sera from HIV-1-infected individuals contain neutralizing antibodies capable of inhibiting cell fusion of infected with uninfected cells and blocking or reducing the infectivity of cell-free virus [126, 127]. A correlation between the presence of such neutralizing antibodies and the clinical status of the patient has been described. Asymptomatic individuals and patients with lymphadenopathy as the only symptom of HIV-1 infection tend to have higher titers than those manifesting AIDS-related complex or AIDS, and often demonstrate an increase in neutralization titer over time [127–130].

Two types of neutralizing antibodies can be distinguished: antibodies that neutralize multiple HIV-1 isolates (so-called group-specific antibodies) and antibodies that display a neutralization capacity restricted to one or few isolates (so-called type-specific antibodies). Group-specific neutralizing antibodies are directed to conserved gp120 conformational epitopes [131–134] and show the capacity to inhibit the binding of gp120 with the cellular receptor CD4 [133–137]. Two human monoclonal antibodies (MoAbs) exhibiting this broad neutralization capacity competed with soluble CD4 (sCD4) for gp120/gp160 binding but not with each other [137, 138]. This suggests that the epitopes on gp120 for these MoAbs directly interact with CD4 but do not overlap [138]. Apparently, more than one discontinuous epitope elicits group-specific neutralizing antibodies. This notion is supported by Chamat et al. [139, 140], who found different spectra of neutralizing activities against HIV-1 isolates among the gp120–CD4 attachment-site-directed antibodies. Group-specific neutralizing antibodies, however, are relatively low-titered and appear only 2–6 months after seroconversion [137, 141, 142].

Type-specific neutralizing antibodies are directed against the third variable envelope region [143–145]. Synthetic peptides derived from this region both bound neutralizing antibodies and elicited neutralizing antibodies upon immunization. In contrast to the group-specific antibodies, type-specific antibodies appeared early in infection and reached high titers to the homologous strain [145, 146]. The V3 region was shown to be highly immunodominant, since most individuals produced antibodies to this epitope [124, 147]. Therefore, the V3 region is often referred to as the principal neutralization domain (PND) [124, 148]. The antibody-binding site of the V3 region is formed by the highly conserved triplet Gly-Pro-Gly (forming the β-turn) and the amino acids that directly flank these amino acids [148–150]. The type specificity of the antibodies is defined by these flanking amino acids that vary from one isolate to another [150]. Antibodies to the V3 region did not interfere with gp120-CD4 binding [151, 152]. In contrast, neutralization via V3 antibodies was established by a postbinding phenomenon [152].

The contribution of V3-directed antibodies to HIV neutralization has recently been investigated [134]. Anti-gp120 antibodies purified by affinity chromatography with an HIV strain SF2 gp120-conjugated Sepharose column were separated in V3-directed and non-V3-directed antibodies (one third of the latter was found to be inhibited by sCD4). V3-directed antibodies were more effective than non-V3-directed antibodies in neutralizing a specific isolate. Together, V3-directed and non-V3 gp120-CD4 attachment-site-directed antibodies represented 66–80% of the total neutralizing activity [140]. The contribution of antibodies directed against other neutralization epitopes [153–158] in natural infection has to be established further.

The in vivo relevance of neutralizing antibodies has been studied. In chimpanzees, inoculation of a mixture of the IIIB isolate and polyclonal IIIB-neutralizing antibodies did not result in HIV-1 infection [159]. Protection from HIV-1 infection has been achieved by passive immunization of a chimpanzee with a high-titered neutralizing immunoglobulin prepared from plasma of healthy individuals [160]. Direct evidence for the protective efficacy of PND antibodies was provided by Emini et al. [161]. They showed that PND monoclonal antibodies passively administered to a chimpanzee mediated protection against challenge with the homologous virus strain. Furthermore, active immunization of chimpanzees with recombinant gp120 but not with gp160 elicited protective immunity against the homologous HIV-1 strain [162]. Although the direct contribution of PND antibodies to this immunity was not established, titers of PND antibodies correlated with protection. In another study, immunization with mixtures of HIV-1 antigens, including synthetic peptides derived from the PND, resulted in sustained high titers of neutralizing antibodies and protection [163].

Conflicting data have been published on the role of antibodies in preventing mother-to-child transmission of HIV-1 in humans. Although a positive association between maternal serum antibodies to the envelope protein gp120 [164] or to V3 peptides [165] and the birth of an uninfected child has been reported, this was not confirmed by others [166–168].

Determinant of Biological Phenotype

A number of studies have provided evidence for the involvement of the V3 domain in cellular tropism. O'Brien et al. [169] demonstrated by constructing recombinant viruses that tropism for mononuclear phagocytes could be determined by a 157-amino-acid region that included the V3 region. In another report, a 94-amino-acid region, again including the entire V3 region, conferred the capacity for productive infection in primary monocytes [170]. The direct involvement of the V3 region itself was demonstrated by Hwang et al. [171], who constructed recombinant viruses chimeric for a 20-amino-acid sequence of the V3 region. Exchange of this sequence from a macrophage-tropic isolate to the T-cell-line-tropic IIIB isolate resulted in a chimeric virus replication competent on macrophages but not on T cell lines. Furthermore, single amino acid changes in the top of the V3 loop were found to be responsible for differences in host range [172] and infectivity on B cell lines [173]. However, two recent studies emphasized the role of regions other than V3 for determining host range and replicative properties of HIV-1 [174, 175].

A role of the V3 loop in cell fusion mediated by the HIV-1 envelope has been implicated by Freed et al. [176, 177]. Mutations at conserved amino acids of the V3 loop blocked or greatly reduced cell fusion without affecting envelope processing or binding to the CD4 receptor molecule. Moreover, the V3 domain has been reported to be a primary determinant for SI capacity and the replication rate [178, 179].

Several models have been proposed for the involvement of the V3 loop in the biological phenotype of the virus. It is likely that the V3 loop is involved after the initial gp120-CD4 binding event. Callahan et al. [180] speculated that interaction of the positively charged V3 loop with a negatively charged region on the cell membrane takes place in order to reduce repulsive charge effects between the two membranes. This region could be a second receptor on the target cell or, for instance, the negatively charged CDR3 region of the CD4 receptor [180]. The CDR3 region has been implicated by several groups as being involved in the infection process after the initial gp120-CD4 binding [181–183]. The fact that V3-derived peptides specifically bound to a CD4 site as distinct from the gp120-binding site provided additional evidence [184]. An alternative model suggested that envelope fusion requires a specific proteolytic activation step and that the V3 domain contains such proteolytic-sensitive sites [185, 186]. Proteolytic cleavage near the top of the loop has been demonstrated to occur in vitro [187, 188]. According to this model, tropism would be determined by the availability of sequence-specific cellular proteases on the target cell.

Impact of Variation of the Third Variable gp120 Region during Natural HIV-1 Infection

An HIV-1 virus population is characterized by the coexistence of numerous virus variants. Genomic diversity of these virus variants as well as continual generation of new variants by mutation allows the viral population to adapt optimally to changing environmental conditions by the selection of variants with different antigenic and biological characteristics.

The Acute Phase of HIV-1 Infection

At the acute HIV-1 infection a homogeneous virus population was found [189, 190; G. Zwart et al., submitted]. The outgrowth of a virus population with limited genomic variation, in the absence of HIV-1-directed antibodies, resembles the loss of variation following virus cultivation observed in vitro. In vivo, rapid outgrowth of the virus variant with the highest relative fitness from the pool of infecting viruses may account for the emergence of a homogeneous population. A second possibility is that a limited number of virus variants succeeds in infection. This second possibility is favored for several reasons. The low transmission rate of HIV-1 infection via sexual intercourse suggests that infection is established by only a limited number of viruses. Also in accordance with this notion is the observation that the variant that grew out was often identical to the virus variant most dominantly present in the virus transmitter [189]. In addition, the early virus population in vivo appears to be even more homogeneous than the virus population obtained after cocultivation.

The acute infection phase is further characterized by high levels of genomic RNA and p24 core antigen, indicative of the rapid initial virus multiplication [190].

The Asymptomatic Phase

After the development of an immunological response against the initially replicated virus variants, titers of genomic RNA decrease and p24 core antigen is normally not detectable in serum. During this stage of the infection, referred to as the asymptomatic phase, genomic RNA often remains detectable in the serum. This argues against a true state of viral latency.

In contrast to the homogeneous population of viral sequences found around the time of antibody conversion, at this stage of the infection the viral population present in serum and PBMCs consists of a heterogeneous mixture of more or less related genotypes. Virus variants with mutations clustered in specific regions are found. Athough no longer found in the peripheral blood at our detection level, there is no evidence that initial virus variants are indeed cleared efficiently from the body. Consequently, the number of virus variants within one individual will be extremely high. Although not all mutations are preserved in consecutive samples and evolution of HIV-1 does not appear to follow one straight evolutionary line, in general genomic divergence of the viral population to the first found population increases over time [190, 191].

The rate at which the total viral population increases in complexity is dependent on numerous variables. First of all, the viral replication rate plays an important role in this. A higher replication rate concurs with more RNA-

DNA transcriptions and consequently with more RT-introduced mutations in the proviral reservoir. Secondly, the dynamics of the virus multiplication has major impact on the rate of evolutionary change. The number of replication cycles and consequently the number of RNA-DNA transcriptions and RT-introduced mutations will correlate inversely with the number of viruses that is transcribed from one proviral genome. Thirdly, fixation of amino acids beneficial for the virus in changed environmental conditions has an important impact on the rate of the virus variability.

Genetic variation of progeny virus may lead to altered antigenic properties compared to the parental virus and escape from immune surveillance. Indirect evidence for the development of antigenic diversity of the V3 region was provided by the observation that most of the nucleotide changes in this region are nonsilent [91, 190, 191]. Immune selection for amino acid mutations that confer such antigenic variation to the virus may explain this positive selection for change. In support of this is the finding that in six children infected with HIV-1 by one batch of infected plasma, the amount of variation between the V3 region and the V3 region of the virus transmitter correlated with the clinical status of the child [191]. Sequences diverged more from the donor sequence among less immunodeficient children than among children who progressed to AIDS. Our (unpublished) observation that sequences from an agammaglobulinemic HIV-1-infected child predominantly showed silent nucleotide mutations corresponds with this. Direct evidence for antigenic variation was provided in a previous study [190], in which we showed that naturally occurring amino acid mutations indeed resulted in antigenic variation. Fixation of an amino acid substitution at gp120 amino acid 308 (KSIHI⇔KSIPI) in the viral populations from two patients reduced the binding affinity of the patients' antibodies and elicited a new specific antibody population. In an extended study we showed that, although the early antibody response against V3 accurately reflected the virus population circulating at the initial phase of infection, this reflection was lost at later stages of infection due to antigenic changes of the virus [G. Zwart et al., submitted].

Although the immune system is able to react to new antigenic variants [190], antibody reactivity to the initially found variant often remains highest. In a group of 47 seroconverters this change in V3 antibody specificity was observed only in three individuals over a period of 5 years [G. Zwart et al., submitted].

Köhler et al. [192] showed that HIV-1 infection establishes a dominance of B cell clones. This clonally restricted anti-HIV-1 immune response can be explained by defects in B cell function caused by the HIV-1 infection. Another explanation is the 'original antigenic sin' mechanism, in which memory B cells are reactivated by a new antigen that is slightly different from the original one without generating a native response. This cross-stimulation of B cell clones leads to the production of high affinity antibodies, primarily against the original virus variant but only weakly cross-reacting to the new variants.

A somewhat different model was proposed by Nowak et al. [193], in which the steady rise in diversity is seen as the direct cause of the immunodeficiency disease. On the one hand, continuing appearance of new antigenic variants enables the virus population to evade elimination by the immune system. On the other, capacity of CD4– cells is debilitated by the virus infection itself. A threshold value is ultimately passed beyond which the immune system is unable to mount an effective immune response against new variants. Both models may be operational.

Assessment of the importance of the described antigenic diversity and antibody reactivity to the PND is complicated by the fact that, not only mutations at positions crucial for antibody binding in the V3 region itself, but also mutations distant from the primary binding site may confer resistance to neutralization [116, 194, N.K.T. Back et al., in preparation]. This escape mechanism has been referred to as the conformational escape mechanism [195]. Although still binding the V3 epitope, these antibodies do not neutralize the viruses. Evidence that the V3 region interacts with other envelope regions has been obtained earlier by Willey et al. [196]. An amino acid change in the V3 region could compensate for a deleterious amino acid change within the second conserved domain of gp120 that markedly reduced HIV-1 infectivity.

The Symptomatic Phase

After a variable duration of the asymptomatic phase, HIV-1 infection becomes symptomatic. This is accompanied by higher levels of plasma virus [197, 198], reflected in our study by a rise in genomic RNA and p24 core antigen [190], and by a decrease in V3-binding antibody titers and low V3 antibody affinity [G. Zwart et al., in preparation]. In 50% of the cases the transition to the symptomatic phase of the HIV-1 infection is preceded by the conversion in phenotype from non-SI to SI viruses [106, 107, 199].

Besides encompassing the PND, V3 plays an important role in this biological feature. We and others [178, 179, 200a] showed that transition from non-SI to SI is associated with changes in the V3 region [190, T.F.W. Wolfs et al., submitted]. Remarkably, the V3 region of SI viruses often shows variation to the preexisting non-SI V3 region at multiple positions. This can mean two things: (1) Mutations on positions associated with the switch from non-SI to SI phenotype [179] are only effective in a certain context and, therefore, accompanied by other amino acid mutations. In this case the altered V3 region may often lead to altered antigenicity of the PND. Evidence for other contributing amino acid changes has recently been obtained by De Jong et al. [200b]. (2) SI viruses originate in a separate compartment distinct from the peripheral blood and follow a separate path of evolution not hampered by the antibody population in the peripheral blood.

In two patients studied we found evidence that changes in the V3 region simultaneously affected antigenic and biological phenotype.

(1) In a transfusion-associated virus transmission case [189], SI isolates were recovered both from virus donor and recipient [201]. Biological clones with SI phenotype recovered from virus donor and recipient displayed characteristic V3 sequences [179] that were very similar to the sequences found in the majority of the clones of the uncultured viral RNA population from the virus donor and in all sequenced clones from the recipient. The highest serological reactivity of the virus donor, however, was against the V3 peptide that was identical to the sequence found in the minority of the clones of the uncultured viral RNA population (and in all clones determined from a sample taken 4 years earlier). No reactivity was found against the V3 peptide that matched best with the sequence of the SI biological clones. In contrast, the highest antibody reactivity shown by the recipient was against this peptide.

(2) In a patient's uncultured DNA and RNA population [T.F.W. Wolfs et al., submitted], V3 regions were found that were associated with SI as well as with non-SI viruses. Serum antibodies of this patient showed reactivity against the non-SI V3 variant. The SI V3 variant, which was not observed in an early sample from this patient, had substitutions at positions crucial for antibody binding (GPGR→GHKR and GPGR→GYKR) and escapes from the existing immune response.

In the early phase of infection, when the immune system is still able to react to new antigens, minimal changes resulting in a (partial) escape from the immune response are of benefit for the virus. Via antigenic sin, such variants will only trigger antibody production to the early variant. Viruses displaying large differences, however, would be recognized by this time. In late infection, when the capacity of the immune system to react to new antigens is minimal, major changes simultaneously affecting antigenic and biological phenotype can survive without a productive activation of the immune system.

Deterioration of the immune system coincided with a decrease in heterogeneity of the present virus population in one longitudinally followed individual, although this is possibly influenced by the administration of zidovudine (AZT). The decrease in heterogeneity can be implicated as a 'clonal' outgrowth of the most replication-competent variant at the stage of infection with the number of infected cells rising dramatically and virulent variants no longer being suppressed by the immune system. The rate at which this predominance of the 'fittest variant' occurs is again dependent on several factors: replication rate, turnover of infected cells and infection grade of the cells. The selection imposed by the immune system, however, will be minimal at this stage. Due to these variables, it will take an individually determined period of time until the more virulent SI variant dominates the non-SI variant in the uncultured PBMCs and serum. Kuiken et al. [200a] found a delay time between the first isolation of an SI virus in vitro and the dominance of SI-associated V3 regions in uncultured material of 3 to 9 months in two patients. Twenty months after conversion for SI phenotype, 35% of the uncultured proviral genomes still harbored the non-SI-associated V3 region [T.F.W. Wolfs et al., submitted]. If indeed only one or very few (randomly selected) infectious units are transmitted, this time delay may account for the observation made by Roos et al. [201] that biological phenotype of virus donor and recipient not always match. It may also partly explain the low number of SI virus infection observed among acutely infected individuals.

Implication of Viral Diversity for Vaccine Development

The described variability of the V3 region together with the type specificity of neutralizing antibodies directed against it poses major problems for the development of a V3-based vaccine. However, the contribution of those antibodies to the overall neutralization capacity of sera is significant. Moreover, the protective value for HIV-1 infection of a V3-directed MoAb has convincingly been demonstrated in a chimpanzee experiment [161]. At a circulating virus-neutralizing antibody titer of 1:320, challenge with the homologous cell-free virus did not result in virus infection. The combination of V3 as PND and important determinant for biological phenotype, however, means that variation of V3 is not unlimited, but subject to certain restrictions (structural or functional). As we observed in our studies and as has been observed also in the greatest available V3 sequence data set [202], some positions within V3 (including the amino acids at the top of the V3 region and those directly flanking the two cysteines) are conserved in most isolates, whereas others are only substituted by a limited number of (conservative) amino acids. Restrictness of variation may be helpful in using V3 peptides for vaccine purposes. Neutralizing antibodies have been raised against one such relatively conserved element of the V3 region [203, 204]. These antibodies were directed against the hexameric amino acid structure at the center of the V3 region. Immunization with the same hexamer (GPGRAF) resulted in antibodies capable of neutralizing viruses containing this hexamer, but not viruses with other amino acids at those positions [203]. Based on these results, the most prevalent pentamers, hexamers and heptamers in the viruses sequenced thus far have been determined in order to predict optimal peptide mixtures for the inducing broadly neutralizing antibodies [205]. A set of 10 hexamers would cover more than 87% of 245 HIV-1 sequences. As we described, intrapatient variation at the early phase of HIV-1 infection is limited. The huge intrapatient variation in late infection, however, makes these figures less promising. Most individuals carried variants that did not match any of the sequences from the 10 most common hexapeptides [206]. The more sequences that were determined for an individual the less chance there was that this sequence set was covered.

If a minimal number of viral infectious units are transmitted at virus infection, protection against HIV-1 infection using peptides derived from the most frequently found naturally occurring V3 regions is a real possibility, provided that high neutralization titers sustain and confer protection from HIV-1 infection. Geographical testing of V3 prevalences is underway (WHO Technical Working Group on HIV Isolation and Characterization). In the long run, however, this type of immunization will increase the prevalence of neutralization-resistant variants. Active or passive immunization based solely on V3-directed antibodies of individuals already infected will only select for escape mutants in the patient and consequently in the population as a whole. Altogether, the variability of the V3 region reemphasizes the importance of studying conserved epitopes that elicit broadly neutralizing antibodies and the use of vaccines based on a combination of both V3-directed and gp120-CD4-blocking antibodies.

References

1 Baltimore D: RNA-dependent DNA polymerase in virions of RNA tumour viruses. Nature 1970;226:1209–1211.
2 Temin HM, Mizutani S: RNA-dependent DNA polymerase in virions of Rous sarcoma virus. Nature 1970;226:1211–1213.
3 Weiner AM, Deininger PL, Efstratiadis A: Nonviral retroposons, genes, pseudogenes, and transposable elements generated by the reverse transcriptase flow of genetic information. Annu Rev Biochem 1986;55:631–661.
4 Boeke JD, Corces VG: Transcription and reverse transcription of retrotransposons. Annu Rev Microbiol 1989;43:403–434.
5 Temin HM: Malignant transformation of cells by viruses. Perspect Biol Med 1970; 14:11–26.
6 Inouye S, Hsu MY, Eagle S, Inouye M: Reverse transcriptase associated with the biosynthesis of the branched RNA-linked msDNA in Myxococcus xanthus. Cell 1989; 56:709–717.
7 Lim D, Maas WK: Reverse transcriptase-dependent synthesis of a covalently linked, branched DNA-RNA compound in E. coli B. Cell 1989;56:891–904.
8 Temin HM: Retrons in bacteria. Nature 1989;339:254–255.
9 Xiong Y, Eickbush TH: Origin and evolution of retro elements based upon their reverse transcriptase sequences. EMBO J 1990;9:3353–3362.
10 Doolittle RF, Feng D-F, Johnson MS, McClure MA: Origins and evolutionary relationships of retroviruses. Q Rev Biol 1989;64:1–30.
11 Xiong Y, Eickbush TH: Similarity of reverse transcriptase-like sequences, of viruses, transposable elements and mitochondrial introns. Mol Biol Evol 1988;5:675–690.
12 Steele PE, Martin MA, Rabson AB, Bryan T, O'Brien SJ: Amplification and chromosomal dispersion of human endogenous retroviral sequences. J Virol 1986; 59:545–550.
13 Shih A, Coutavas EE, Rush MG: Evolutionary implications of primate endogenous retroviruses. Virology 1991;182:495–502.
14 Horwitz MS, Boyce-Jacino MT, Faras AJ: Novel human homogenous sequences related to human immunodeficiency virus type 1. J Virol 1992;66:2170–2179.
15 Shih A, Misra R, Rush MG: Detection of multiple, novel reverse transcriptase coding sequences in human nucleic acids: Relation to primate retroviruses. J Virol 1989;63:64–75.
16 CDC: Pneumocystis pneumonia – Los Angeles. MMWR 1981;30:250–252.
17 Gottlieb MS, Schroff R, Schanker HM, Weisman JD, Fan PT, Wolf RA, Saxon A: Pneumocystis carinii pneumonia and mucosal candidiasis in previously healthy homosexual men. N Engl J Med 1981;305:1426–1431.
18 Masur H, Michelis MA, Greene JB, Onorato I, vande Stouwe RA, Holzman RS, Wormser G, Brettman L, Lange M, Murray HW, Cunningham-Rundles S: An outbreak of community-acquired Pneumocystis carinii pneumonia. N Engl J Med 1981;305:1431–1438.

19 Siegal FP, Lopez C, Hammer GS, Brown AE, Kornfeld SJ, Gold J, Hassett J, Hirschman SZ, Cunningham-Rundles C, Adelsberg BR, Parham DM, Siegal M, Cunningham-Rundles S, Armstrong D: Severe acquired immunodeficiency in male homosexuals, manifested by chronic perianal ulcerative herpes simplex lesions. N Engl J Med 1981;305:1439–1444.
20 CDC: Pneumocystis carinii pneumonia among persons with hemophilia A. MMWR 1982;31:365–367.
21 CDC: Possible transfusion-associated acquired immune deficiency syndrome (AIDS) - California. MMWR 1982;31:652–654.
22 CDC: Update on Kaposi's sarcoma and opportunistic infections in previously healthy persons - United States. MMWR 1982;31:294–301.
23 CDC: Unexplained immunodeficiency and opportunistic infections in infants - New York, New Jersey, California. MMWR 1982;31:665–667.
24 CDC: Immunodeficiency among female sexual partners of males with acquired immune deficiency syndrome (AIDS) - New York. MMWR 1983;31:697–698.
25 Barré-Sinoussi F, Chermann JC, Rey F, Nugeyre MT, Chamaret S, Gruest J, Dauguet C, Axler-Blin C, Vézinet-Brun F, Rouzioux C, Rozenbaum W, Montagnier L: Isolation of a T-lymphotropic retrovirus from a patient at risk for acquired immune deficiency syndrome (AIDS). Science 1983;220:868–871.
26 Coffin J, Haase A, Levy JA, Montagnier L, Oroszlan S, Teich N, Temin H, Toyoshima K, Varmus H, Vogt P, Weiss R: What to call the AIDS virus? Nature 1986;321:10.
27 Blattner WA: HIV epidemiology: Past, present, and future. FASEB J 1991;5:2340–2348.
28 Clavel F, Guétard D, Brun-Vezinet F, Chamaret S, Rey M-A, Santos-Ferreira MO, Laurent AG, Dauguet C, Katlama C, Rouzioux C, Klatzmann D, Champalimaud JL, Montagnier L: Isolation of a new human retrovirus from West African patients with AIDS. Science 1986;233:343–346.
29 Clavel F, Guyader M, Guétard D, Sallé M, Montagnier L, Alizon M: Molecular cloning and polymorphism of the human immune deficiency virus type 2. Nature 1986;324:691–695.
30 Letvin NL, Eaton KA, Aldrich WR, Sehgal PK, Blake BJ, Schlossman SF, King NW, Hunt RD: Acquired immunodeficiency syndrome in a colony of macaque monkeys. Proc Natl Acad Sci USA 1983;80:2718–2722.
31 Daniel MD, Letvin NL, King NW, Kannagi M, Sehgal PK, Hunt RD, Kanki PJ, Essex M, Desrosiers RC: Isolation of T-cell tropic HTLV-III-like retrovirus from macaques. Science 1985;228:1201–1204.
32 Kanki PJ, McLane MF, King NW Jr, Letvin NL, Hunt RD, Sehgal P, Daniel MD, Desrosiers RC, Essex M: Serological identification and characterization of a macaque T-lymphotropic retrovirus closely related to HTLV-III. Science 1985;228: 1199–1201.
33 Lowenstine LJ, Pedersen NC, Higgins J, Pallis KC, Uyeda A, Marx P, Lerche NW, Munn RJ, Gardner MB: Seroepidemiologic survey of captive old-world primates for antibodies to human and simian retroviruses, and isolation of a lentivirus from sooty mangabeys (Cercobus atys). Int J Cancer 1986;38:563–574.
34 Ohta Y, Masuda T, Tsujimoto H, Ishikawa K, Kodama T, Morikawa S, Nakai M, Honjo S, Hayami M: Isolation of simian immunodeficiency virus from African green

monkeys and seroepidemiologic survey of the virus in various non-human primates. Int J Cancer 1988;41:115–122.

35 Ratner L, Haseltine W, Patarca R, Livak KJ, Starcich B, Josephs SF, Doran ER, Rafalski JA, Whitehorn EA, Baumeister K, Ivanoff L, Petteway SR Jr, Pearson ML, Lautenberger JA, Papas TS, Ghrayeb J, Chang NT, Gallo RC, Wong-Staal F: Complete nucleotide sequence of the AIDS virus, HTLV-III. Nature 1985;313: 277–284.

36 Guyader M, Emerman M, Sonigo P, Clavel F, Montagnier L, Alizon M: Genome organization and transactivation of the human immunodeficiency virus type 2. Nature 1987;326:662–669.

37 Franchini G, Gurgo C, Guo H-G, Gallo RC, Collalti E, Fargnoli KA, Hall LF, Wong-Staal F, Reitz MS Jr: Sequence of simian immunodeficiency virus and its relationship to the human immunodeficiency viruses. Nature 1987;328:539–543.

38 Chakrabarti L, Guyader M, Alizon M, Daniel MD, Desrosiers RC, Tiollais P, Sonigo P: Sequence of simian immunodeficiency virus from macaque and its relationship to other human and simian retroviruses. Nature 1987;328:543–547.

39 Hirsch V, Riedel N, Mullins JI: The genome organization of STLV-3 is similar to that of the AIDS virus except for a truncated transmembrane protein. Cell 1987; 49:307–319.

40 Murphey-Corb M, Martin LN, Rangan SRS, Baskin GB, Gormus BJ, Wolf RH, Andes WA, West M, Montelaro RC: Isolation of an HTLV-III-related retrovirus from macaques with simian AIDS and its possible origin in asymptomatic mangabeys. Nature 1986;321:435–437.

41 Fultz PN, McClure HM, Swenson CR, McGrath CR, Brodie A, Getchell JP, Jensen FC, Anderson DC, Broderson JR, Francis DP: Persistent infection of chimpanzees with human T-lymphotropic virus type III/lymphadenopathy-associated virus: A potential model for acquired immunodeficiency syndrome. J Virol 1986;58:116–124.

42 Hirsch VM, Olmsted RA, Murphey-Corb M, Purcell RH, Johnson PR: An African primate lentivirus (SIVsm) closely related to HIV-2. Nature 1989;339:389–392.

43 Peeters M, Honoré C, Huet T, Bedjabaga L, Ossari S, Bussi P, Cooper RW, Delaporte E: Isolation and partial characterization of an HIV-related virus occurring naturally in chimpanzees in Gabon. AIDS 1989;3:625–630.

44 Huet T, Cheynier R, Meyerhans A, Roelants G, Wain-Hobson S: Genetic organization of a chimpanzee lentivirus related to HIV-1. Nature 1990;345:356–359.

45 Johnson PR, Fomsgaard A, Allan J, Gravell M, London WT, Olmsted RA, Hirsch VM: Simian immunodeficiency viruses from African green monkeys display unusual genetic diversity. J Virol 1990;64:1086–1092.

46 Hendry RM, Wells MA, Phelan MA, Schneider AL, Epstein JS, Quinnan GV: Antibodies to simian immunodeficiency virus in African green monkeys in Africa in 1957–62. Lancet 1986;ii:455.

47 Fomsgaard A, Hirsch VM, Allan JS, Johnson PR: A highly divergent proviral DNA clone of SIV from a distinct species of African green monkey. Virology 1991;182: 397–402.

48 Tsujimoto H, Cooper RW, Kodama T, Fukasawa M, Miura T, Ohta Y, Ishikawa K-I, Nakai M, Frost E, Roelants GE, Roffi J, Hayami M: Isolation and characterization of simian immunodeficiency virus from mandrils in Africa and its relationship to other human and simian immunodeficiency viruses. J Virol 1988;62:4044–4050.

49 Sharp PM, Li W-H: Understanding the origins of AIDS viruses. Nature 1988; 336:315.
50 Yokoyama S, Chung L, Gobojori T: Molecular evolution of the human immunodeficiency and related viruses. Mol Biol Evol 1988;5:237–251.
51 Khan AS, Galvin TA, Lowenstine LJ, Jennings MB, Gardner MB, Buckler CE: A highly divergent simian immunodeficiency virus (SIVstm) recovered from stored stump-tailed macaque tissues. J Virol 1991;65:7061–7065.
52 Doolittle RF: The simian-human connection. Nature 1989;339:338–339.
53 Li W-H, Tanimura M, Sharp PM: Rates and dates of divergence between AIDS virus nucleotide sequences. Mol Biol Evol 1988;5:313–330.
54 Nahmias AJ, Weiss J, Yao X, Lee F, Kodsi R, Schanfield M, Matthews T, Bolognesi D, Durack D, Motulsky A, Kanki P, Essex M: Evidence for human infection with an HTLV III/LAV-like virus in Central Africa, 1959. Lancet 1986;i:1279–1280.
55 Eigen M, Nieselt-Struwe K: How old is the immunodeficiency virus? AIDS 1990; 4(suppl 1):S85–S93.
56 Roberts JD, Bebenek K, Kunkel TA: The accuracy of reverse transcriptase from HIV-1. Science 1988;242:1171–1173.
57 Preston BD, Poiesz BJ, Loeb LA: Fidelity of HIV-1 reverse transcriptase. Science 1988;242:1168–1171.
58 Ji J, Loeb LA: Fidelity of HIV-1 reverse transcriptase copying RNA in vitro. Biochemistry, in press.
59 Ricchetti M, Buc H: Reverse transcriptases and genomic variability: The accuracy of DNA replication is enzyme specific and sequence dependent. EMBO J 1990;9:1583–1593.
60 Weber J, Grosse F: Fidelity of human immunodeficiency virus type I reverse transcriptase in copying natural DNA. Nucleic Acids Res 1989;17:1379–1393.
61 Roberts JD, Preston BD, Johnston LA, Soni A, Loeb LA, Kunkel TA: Fidelity of two retroviral reverse transcriptases during DNA-dependent DNA synthesis in vitro. Mol Cell Biol 1989;9:469–476.
62 Hu W-S, Temin HM: Genetic consequences of packaging two RNA genomes in one retroviral particle: Pseudodiploidy and high rate of genetic recombination. Proc Natl Acad Sci USA 1990;87:1556–1560.
63 Coffin JM: Structure, replication, and recombination of retrovirus genomes: Some unifying hypotheses. J Gen Virol 1979;42:1–26.
64 Junghans RP, Boone LR, Skalka AM: Retroviral DNA H structures: Displacement-assimilation model of recombination. Cell 1982;30:53–62.
65 Hu W-S, Temin HM: Retroviral recombination and reverse transcription. Science 1990;250:1227–1233.
66 Shaw GM, Hahn BH, Arya SK, Groopman JE, Gallo RC, Wong-Staal F: Molecular characterization of human T-cell leukemia (lymphotropic) virus type III in the acquired immune deficiency syndrome. Science 1984;226:1165–1171.
67 Shaw GM, Harper ME, Hahn BH, Epstein LG, Gajdusek DC, Price RW, Navia BA, Petito CK, O'Hara CJ, Groopman JE, Cho E-S, Oleske JM, Wong-Staal F, Gallo RC: HTLV-III infection in brains of children and adults with AIDS encephalopathy. Science 1985;227:177–182.
68 Hahn BH, Shaw GM, Arya SK, Popovic M, Gallo RC, Wong-Staal F: Molecular cloning and characterization of the HTLV-III virus associated with AIDS. Nature 1984;312:166–169.

69 Hahn BH, Gonda MA, Shaw GM, Popovic M, Hoxie JA, Gallo RC, Wong-Staal F: Genomic diversity of the acquired immune deficiency syndrome virus HTLV-III: Different viruses exhibit greatest divergence in their envelope genes. Proc Natl Acad Sci USA 1985;82:4813–4817.
70 Wong-Staal F, Shaw GM, Hahn BH, Salahuddin SZ, Popovic M, Markham P, Redfield R, Gallo RC: Genomic diversity of human T-lymphotropic virus type III (HTLV-III). Science 1985;229:759–762.
71 Wain-Hobson S, Sonigo P, Danos O, Cole S, Alizon M: Nucleotide sequence of the AIDS virus, LAV. Cell 1985;40:9–17.
72 Sanchez-Pescador R, Power MD, Barr PJ, Steimer KS, Stempien MM, Brown-Shimer SL, Gee WW, Renard A, Randolph A, Levy JA, Dina D, Luciw PA: Nucleotide sequence and expression of an AIDS-associated retrovirus (ARV-2). Science 1985;227:484–492.
73 Alizon M, Wain-Hobson S, Montagnier L, Sonigo P: Genetic variability of the AIDS virus: Nucleotide sequence analysis of two isolates from African patients. Cell 1986; 46:63–74.
74 Willey RL, Rutledge RA, Dias S, Folks T, Theodore T, Buckler CE, Martin MA: Identification of conserved and divergent domains within the envelope gene of the acquired immunodeficiency syndrome retrovirus. Proc Natl Acad Sci USA 1986; 83:5038–5042.
75 Starcich BR, Hahn BH, Shaw GM, McNeely PD, Modrow S, Wolf H, Parks ES, Parks WP, Josephs SF, Gallo RC, Wong-Staal F: Identification and characterization of conserved and variable regions in the envelope gene of HTLV-III/LAV, the retrovirus of AIDS. Cell 1986;45:637–648.
76 Modrow S, Hahn BH, Shaw GM, Gallo RC, Wong-Staal F, Wolf H: Computer-assisted analysis of envelope protein sequences of seven human immunodeficiency virus isolates: Prediction of antigenic epitopes in conserved and variable regions. J Virol 1987;61:570–578.
77 Hahn BH, Shaw GM, Taylor ME, Redfield RR, Markham PD, Salahuddin SZ, Wong-Staal F, Gallo RC, Parks ES, Parks WP: Genetic variation in HTLV-III/LAV over time in patients with AIDS or at risk for AIDS. Science 1986;232:1548–1553.
78 Saag MS, Hahn BH, Gibbons J, Li Y, Parks ES, Parks WP, Shaw GM: Extensive variation of human immunodeficiency virus type 1 in vivo. Nature 1988;334:440–444.
79 Fisher AG, Ensoli B, Looney D, Rose A, Gallo RC, Saag MS, Shaw GM, Hahn BH, Wong-Staal F: Biologically diverse molecular variants within a single HIV-1 isolate. Nature 1988;334:444–447.
80 Saiki RK, Scharf S, Faloona F, Mullis KB, Horn GT, Erlich HA, Arnheim N: Enzymatic amplification of β-globin genomic sequences and restriction site analysis for diagnosis of sickle cell anemia. Science 1985;230:1350–1354.
81 Goodenow M, Huet T, Saurin W, Kwok S, Sninsky J, Wain-Hobson S: HIV-1 isolates are rapidly evolving quasispecies: Evidence for viral mixtures and preferred nucleotide substitutions. J AIDS 1989;2:344–352.
82 Eigen M: Self-organization of matter and the evolution of biological macromolecules. Naturwissenschaften 1971;58:465–523.
83 Eigen M, McCaskill J, Schuster P: Molecular quasi-species. J Phys Chem 1988;92: 6881–6891.

84 Meyerhans A, Cheynier R, Albert J, Seth M, Kwok S, Sninsky J, Morfeldt-Manson L, Åsjö B, Wain-Hobson S: Temporal fluctuations in HIV quasispecies in vivo are not reflected by sequential HIV isolations. Cell 1989;58:901–910.
85 Delassus S, Cheynier R, Wain-Hobson S: Evolution of human immunodeficiency virus type 1 nef and long terminal repeat sequences over 4 years in vivo and in vitro. J Virol 1991;65:225–231.
86 Vartanian J-P, Meyerhans A, Åsjö B, Wain-Hobson S: Selection, recombination and G→A hypermutation of human immunodeficiency virus type 1 genomes. J Virol 1991;65:1779–1788.
87 Pang S, Vinters HV, Akashi T, O'Brien WA, Chen ISY: HIV-1 env sequence variation in brain tissue of patients with AIDS-related neurologic disease. J AIDS 1991;4:1082–1092.
88 Martins LP, Chenciner N, Åsjö B, Meyerhans A, Wain-Hobson S: Independent fluctuation of human immunodeficiency virus type 1 rev and gp41 quasispecies in vivo. J Virol 1991;65:4502–4507.
89 Kusumi K, Conway B, Cunningham S, Berson A, Evans C, Iversen AKN, Colvin D, Gallo MV, Coutre S, Shpaer EG, Faulkner DV, de Ronde A, Volkman S, Williams C, Hirsch MS, Mullins JI: Human immunodeficiency virus type 1 envelope gene structure and diversity in vivo and after cocultivation in vitro. J Virol 1992;66:875–885.
90 Balfe P, Simmonds P, Ludlam A, Bishop JO, Leigh Brown AJ: Concurrent evolution of human immunodeficiency virus type 1 in patients infected from the same source: Rate of sequence change and low frequency of inactivating mutations. J Virol 1990; 64:6221–6233.
91 Simmonds P, Balfe P, Ludlam CA, Bishop JO, Leigh Brown AJ: Analysis of sequence diversity in hypervariable regions of the external glycoprotein of human immunodeficiency virus type 1. J Virol 1990;64:5840–5850.
92 Burger H, Gibbs RA, Nguyen PN, Flaherty K, Gulla J, Belman A, Weiser B: HIV-1 transmission within a family: Generation of viral heterogeneity correlates with duration of infection; in Brown F, Chanock RM, Ginsberg HS, Lerner RA (eds): Vaccines 90. New York, Cold Spring Harbor Laboratory Press, 1990.
93 McNearney T, Westervelt P, Thielan BJ, Trowbridge DB, Garcia J, Whittier R, Ratner L: Limited sequence heterogeneity among biologically distinct human immunodeficiency virus type 1 isolates from individuals involved in a clustered infectious outbreak. Proc Natl Acad Sci USA 1990;87:1917–1921.
94 Kleim JP, Ackermann A, Brackmann HH, Gahr M, Schneweis KE: Epidemiologically closely related viruses from hemophilia B patients display high homology in two hypervariable regions of the HIV-1 env gene. AIDS Res Hum Retroviruses 1991; 7:417–421.
95 Epstein LG, Kuiken C, Blumberg BM, Hartman S, Sharer LR, Clement M, Goudsmit J: HIV-1 V3 domain variation in brain and spleen of children with AIDS: Tissue-specific evolution within host-determined quasispecies. Virology 1991;180: 583–590.
96 Simmonds P, Zhang LQ, McOmish F, Balfe P, Ludlam CA, Leigh Brown AJ: Discontinuous sequence change of human immunodeficiency virus (HIV) type 1 env sequences in plasma viral and lymphocyte-associated proviral populations in vivo: Implications for models of HIV pathogenesis. J Virol 1991;65:6266–6276.
97 Anand R, Siegal F, Reed C, Cheung T, Forlenza S, Moore J: Non-cytocidal natural

variants of human immunodeficiency virus isolated from AIDS patients with neurological disorders. Lancet 1987;ii:234–238.

98 Åsjö B, Morfeldt-Månson L, Albert J, Biberfeld G, Karlsson A, Lidman K, Fenyö EM: Replicative capacity of human immunodeficiency virus from patients with varying severity of HIV infection. Lancet 1986;ii:660–662.

99 Dahl K, Martin K, Miller G: Differences among human immunodeficiency virus strains in their capacities to induce cytolysis or persistent infection of a lymphoblastoid cell line immortalized by Epstein-Barr virus. J Virol 1987;61:1602–1608.

100 Evans LA, McHugh TM, Stites DP, Levy JA: Differential ability of human immunodeficiency virus isolates to productively infect human cells. J Immunol 1987;138:3415–3418.

101 Briesen von H, Becker WB, Henco K, Helm EB, Gelderblom HR, Brede HD, Rübsamen-Waigmann H: Isolation frequency and growth properties of HIV variants: Multiple simultaneous variants in a patient demonstrated by molecular cloning. J Med Virol 1987;23:51–66.

102 Levy JA, Shimabukuro J, McHugh T, Casavant C, Stites D, Oshiro L: AIDS-associated retroviruses (ARV) can productively infect other cells besides human T helper cells. Virology 1985;147:441–448.

103 Tersmette M, Goede de REY, Al BJM, Winkel IN, Coutinho RA, Cuypers HThM, Huisman JG, Miedema F: Differential syncytium-inducing capacity of HIV isolates: Frequent detection of syncytium-inducing isolates in patients with AIDS and ARC. J Virol 1988;62:2026–2032.

104 Fenyö EM, Morfeld-Manson L, Chiodi F, Lind B, von Gegerfelt A, Albert J, Olausson E, Åsjö B: Distinct replicative and cytopathic characteristics of human immunodeficiency virus isolates. J Virol 1988;62:4414–4419.

105 Cheng-Mayer C, Seto D, Tateno M, Levy JA: Biologic features of HIV-1 that correlate with virulence in the host. Science 1988;240:80–82.

106 Tersmette M, Gruters RA, de Wolf F, de Goede REY, Lange JMA, Schellekens PTH, Goudsmit J, Huisman JG, Miedema F: Evidence for a role of virulent human immunodeficiency virus (HIV) variants in the pathogenesis of acquired immunodeficiency syndrome: Studies on sequential HIV isolates. J Virol 1989;63:2118–2125.

107 Tersmette M, Lange JMA, de Goede REY, de Wolf F, Eeftinck Schattenkerk JKM, Schellekens PTH, Coutinho RA, Huisman JG, Goudsmit J, Miedema F: Association between biological properties of human immunodeficiency virus variants and risk for AIDS and AIDS mortality. Lancet 1989;i:983–985.

108 Meltzer MS, Skillman DR, Hoover DL, Hanson BD, Turpin JA, Kalter DC, Gendelman HE: HIV and the immune system: Macrophages and the human immunodeficiency virus. Immunol Today 1990;11:217–223.

109 Schuitemaker H, Kootstra NA, de Goede REY, de Wolf F, Miedema F, Tersmette M: Monocytotropic human immunodeficiency virus type 1 (HIV-1) variants detectable in all stages of HIV-1 infection lack T-cell line tropism and syncytium-inducing ability in primary T-cell culture. J Virol 1991;65:356–363.

110 Tersmette M, Miedema F: Interactions between HIV and the host immune system in the pathogenesis of AIDS. AIDS 1990;4(suppl 1):S57–S66.

111 Groenink M, Fouchier RAM, de Goede REY, de Wolf F, Gruters RA, Cuypers HThM, Huisman HG, Tersmette M: Phenotypic heterogeneity in a panel of infec-

tious molecular human immunodeficiency virus type 1 clones derived from a single individual. J Virol 1991;65:1968–1975.
112 Schuitemaker H, Koot M, Kootstra NA, Dercksen MW, de Goede REY, van Steenwijk RP, Lange JMA, Eeftinck Schattenkerk JKM, Miedema F, Tersmette M: Biological phenotype of human immunodeficiency virus type 1 clones at different stages of infection: Progression of disease is associated with a shift from monocytotropic to T-cell-tropic virus populations. J Virol 1992;66:1354–1360.
113 Cheng-Mayer C, Homsy J, Evans LA, Levy JA: Identification of human immunodeficiency virus subtypes with distinct patterns of sensitivity to serum neutralization. Proc Natl Acad Sci USA 1988;85:2815–2819.
114 Reitz MS Jr, Wilson C, Naugle C, Gallo RC, Robert-Guroff M: Generation of a neutralization-resistant variant of HIV-1 is due to selection for a point mutation in the envelope gene. Cell 1988;54:57–63.
115 McKeating JA, Gow J, Goudsmit J, Pearl LH, Mulder C, Weiss RA: Characterization of HIV-1 neutralization escape mutants. AIDS 1989;3:777–784.
116 Nara PL, Smit SL, Dunlop N, Hatch W, Merges M, Waters D, Kelliher J, Gallo RC, Fischinger PJ, Goudsmit J: Emergence of viruses resistant to neutralization by V3-specific antibodies in experimental human immunodeficiency virus type 1 IIIB infection of chimpanzees. J Virol 1990;64:3779–3791.
117 Albert J, Abrahamsson B, Nagy K, Aurelius E, Gaines SH, Nyström G, Fenyö EM: Rapid development of isolate-specific neutralizing antibodies after primary HIV-1 infection and consequent emergence of virus variants which resist neutralization by autologous sera. AIDS 1990;4:107–112.
118 Narayan O, Griffin DE, Chase J: Antigenic shift of visna virus in persistently infected sheep. Science 1977;197:376–378.
119 Clements JE, Pederson FS, Narayan O, Haseltine WA: Genomic changes associated with antigenic variation of visna virus during a persistent infection. Proc Natl Acad Sci USA 1980;77:4454–4458.
120 Narayan O, Clements JE, Griffin DE, Wolinsky JS: Neutralizing antibody spectrum determines the antigenic profiles of emerging mutants of visna virus. Infect Immun 1981;32:1045–1050.
121 Kono Y, Kobayashi K, Fukunaga Y: Antigenic drift of equine infectious anemia virus in chronically infected horses. Arch Ges Virusforsch 1973;41:1–10.
122 Salinovich O, Payne SL, Montelero RC, Hussam KA, Issel CJ, Schnorr KL: Rapid emergence of novel antigenic and genetic variants of EIAV during persistent infection. J Virol 1986;57:71–80.
123 Leonard CK, Spellman MW, Riddle L, Harris RJ, Thomas JN, Gregory TJ: Assignment of intrachain disulfide bonds and characterization of potential glycosylation sites of the type 1 recombinant human immunodeficiency virus envelope glycoprotein (gp120) expressed in Chinese hamster ovary cells. J Biol Chem 1990; 265:10373–10382.
124 LaRosa GJ, Davide JP, Weinhold K, Waterbury JA, Profy AT, Lewis JA, Langlois AJ, Dreesman GR, Boswell RN, Shadduck P, Holley LH, Karplus M, Bolognesi DP, Matthews TJ, Emini EA, Putney SD: Conserved sequence and structural elements in the HIV-1 principal neutralizing determinant. Science 1990;249:932–935.
125 Chandrasekhar K, Profy AT, Dyson HJ: Solution conformational preferences of immunogenic peptides derived from the principal neutralizing determinant of HIV-1 envelope glycoprotein gp120. Biochemistry 1991;30:9187–9194.

126 Weiss RA, Clapham PR, Cheingsong-Popov R, Dalgleish AG, Carne CA, Weller IVD, Tedder RS: Neutralization of human T-lymphotropic virus type III by sera of AIDS and AIDS-risk patients. Nature 1985;316:69–72.
127 Robert-Guroff M, Brown M, Gallo RC: HTLV-III-neutralizing antibodies in patients with AIDS and AIDS-related complex. Nature 1985;316:72.
128 Weber JN, Clapham PR, Weiss RA, Parker D, Roberts C, Duncan J, Weller I, Carne Ch, Tedder RS, Pinching AJ, Cheingsong-Popov R: Human immunodeficiency virus infection in two cohorts of homosexual men: Neutralising sera and association of anti-gag antibody with prognosis. Lancet 1987;i:119–122.
129 Robert-Guroff M, Goedert JJ, Naugle CJ, Jennings AM, Blattner WA, Gallo RC: Spectrum of HIV-1 neutralizing antibodies in a cohort of homosexual men: Results of a 6 year prospective study. AIDS Res Hum Retroviruses 1988;4:343.
130 Ranki A, Weiss SH, Valle S-L, Antonen J, Krohn KJE: Neutralizing antibodies in HIV (HTLV-III) infection: Correlation with clinical outcome and antibody response against different viral proteins. Clin Exp Immunol 1987;69:231–239.
131 Haigwood NL, Barker CB, Higgins KW, Skiles PV, Moore GK, Mann KA, Lee DR, Eichberg JW, Steimer KS: Evidence for neutralizing antibodies directed against conformational epitopes of HIV-1 gp120; in Brown F, Chanock RM, Ginsberg HS, Lerner RA (eds): Vaccines 90. New York, Cold Spring Harbor Laboratory Press, 1990, pp 313–319.
132 Profy AT, Salinas PA, Eckler LI, Dunlop NM, Nara PL, Putney SD: Epitopes recognized by the neutralizing antibodies of an HIV-1-infected individual. J Immunol 1990;144:4641–4647.
133 Steimer KS, Scandella CJ, Skiles PV, Haigwood NL: Neutralization of divergent HIV-1 isolates by conformation-dependent human antibodies to gp120. Science 1991;254:105–108.
134 Kang C-Y, Nara P, Chamat S, Caralli V, Ryskamp T, Haigwood N, Newman R, Koehler H: Evidence for non-V3-specific neutralizing antibodies that interfere with gp120/CD4 binding in human immunodeficiency virus 1-infected humans. Proc Natl Acad Sci USA 1991;88:6171–6175.
135 Skinner MA, Langlois AJ, McDanal CB, McDougal S, Bolognesi DP, Matthews TJ: Neutralizing antibodies to an immunodominant envelope sequence do not prevent gp120 binding to CD4. J Virol 1988;62:4195–4200.
136 Back NKT, Thiriart C, Delers A, Ramautarsing C, Bruck C, Goudsmit J: Association of antibodies blocking HIV-1 gp160-sCD4 attachment with virus neutralizing activity in human sera. J Med Virol 1990;31:200–208.
137 Ho DD, McKeating JA, Li XL, Moudgil T, Daar ES, Sun NC, Robinson JE: Conformational epitope on gp120 important in CD4 binding and human immunodeficiency virus type 1 neutralization identified by a human monoclonal antibody. J Virol 1991;65:489–493.
138 Ho DD, Fung MSC, Cao Y, Li XL, Sun C, Chang TW, Sun N-C: Another discontinuous epitope on the glycoprotein gp120 that is important in human immunodeficiency virus type 1 neutralization is identified by a monoclonal antibody. Proc Natl Acad Sci USA 1991;88:8949–8952.
139 Chamat S, Nara P, Berquist L, Whalley A, Morrow WJW, Köhler H, Kang C-Y: Two major groups of neutralizing anti-gp120 antibodies exist in HIV-infected individuals: Evidence for epitope diversity around the CD4 attachment site (abstract). Modern Approaches to New Vaccines. Cold Spring Harbor Laboratory, 1991, p 26.

140 Chamat S, Nara P, Berquist L, Whalley A, Morrow WJW, Köhler H, Kang C-Y: Different neutralizing antibody activities indicate epitope diversity around the CD4 attachment site of gp120. Science, in press.
141 Goudsmit J, Thiriart C, Smit L, Bruck C, Gibbs CJ Jr: Temporal development of cross-neutralization between HTLV-III B and HTLV-III RF in experimentally infected chimpanzees. Vaccine 1988;6:229–232.
142 Nara PL, Robey WG, Arthur LO, Asher DM, Wolff AV, Gibbs CJ Jr, Gajdusek DC, Fischinger PJ: Persistent infection of chimpanzees with human immunodeficiency virus: Serological responses and properties of reisolated viruses. J Virol 1987;61: 3173–3180.
143 Palker TJ, Clark ME, Langlois AJ, Matthews TJ, Weinhold KJ, Randall RR, Bolognesi DP, Haynes BF: Type-specific neutralization of the human immunodeficiency virus with antibodies to env-encoded synthetic peptides. Proc Natl Acad Sci USA 1988;85:1932–1936.
144 Rusche JR, Javaherian K, McDanal C, Petro J, Lynn DL, Grimaila R, Langlois A, Gallo RC, Arthur LO, Fischinger PJ, Bolognesi DP, Putney SD, Matthews TJ: Antibodies that inhibit fusion of human immunodeficiency virus-infected cells bind a 24-amino acid sequence of the viral envelope, gp120. Proc Natl Acad Sci USA 1988;85:3198–3202.
145 Goudsmit J, Debouck C, Meloen RH, Smit L, Bakker M, Asher DM, Wolff AV, Gibbs CJ Jr, Gajdusek DC: Human immunodeficiency virus type 1 neutralization epitope with conserved architecture elicits early type-specific antibodies in experimentally infected chimpanzees. Proc Natl Acad Sci USA 1988;85:4478–4482.
146 Nara PL, Robey WG, Pyle SW, Hatch WC, Dunlop NM, Bess JW Jr, Kelliher JC, Arthur LO, Fischinger PJ: Purified envelope glycoproteins from human immunodeficiency virus type 1 variants induce individual, type-specific neutralizing antibodies. J Virol 1988;62:2622–2628.
147 Zwart G, Langedijk H, van der Hoek L, de Jong J-J, Wolfs TFW, Ramautarsing C, Bakker M, de Ronde A, Goudsmit J: Immunodominance and antigenic variation of the principal neutralization domain of HIV-1. Virology 1991;181:481–489.
148 Javaherian K, Langlois AJ, McDanal C, Ross KL, Eckler LI, Jellis CL, Profy AT, Rusche JR, Bolognesi DP, Putney SD, Matthews TJ: Principal neutralizing domain of the human immunodeficiency virus type 1 envelope protein. Proc Natl Acad Sci USA 1989;86:6768–6772.
149 Goudsmit J, Boucher CAB, Meloen RH, Epstein LG, Smit L, van der Hoek L, Bakker M: Human antibody response to a strain-specific HIV-1 gp120 epitope associated with cell fusion inhibition. AIDS 1988;2:157–164.
150 Meloen RH, Liskamp RM, Goudsmit J: Specificity and function of the individual amino acids of an important determinant of human immunodeficiency virus type 1 that induces neutralizing activity. J Gen Virol 1989;70:1505–1512.
151 Linsley PS, Ledbetter JA, Kinney-Thomas E, Hu S-L: Effects of anti-gp120 monoclonal antibodies on CD4 receptor binding by the env protein of human immunodeficiency virus type 1. J Virol 1988;62:3695–3702.
152 Nara PL: HIV-1 neutralization: Evidence for rapid, binding/postbinding neutralization from infected humans, chimpanzees, and gp120-vaccinated animals; in Lerner RA, Ginsberg H, Chanock RM, Brown F (eds): Vaccines 89. New York, Cold Spring Harbor Laboratory Press, 1989, pp 137–144.

153 Sun NC, Ho DD, Sun CRY, Liou RS, Gordon W, Fung MSC, Li XL, Ting R, Lee T-H, Chang NT, Chang TW: Generation and characterization of monoclonal antibodies to the putative CD4-binding domain of human immunodeficiency virus type 1 gp120. J Virol 1989;63:3579–3585.

154 Sarin PS, Sun DK, Thornton AH, Naylor PH, Goldstein AL: Neutralisation of HTLV-III/LAV replication by antiserum to thymosine alpha 1. Science 1986;232: 1135–1137.

155 Ho DD, Sarngadharan MG, Hirsch MS, Schooley RT, Rota TR, Kennedy RC, Chanh TC, Sato VL: Human immunodeficiency virus neutralizing antibodies recognize several conserved domains on the envelope glycoproteins. J Virol 1987;61: 2024–2028.

156 Ho DD, Kaplan JC, Rackauskas IE, Gurney ME: Second conserved domain of gp120 is important for HIV infectivity and antibody neutralization. Science 1988;239: 1021–1023.

157 Dalgleish AG, Chanh TC, Kennedy RC, Kanda P, Clapham PR, Weiss RA: Neutralization of diverse HIV-1 strains by monoclonal antibodies raised against a gp41 synthetic peptide. Virology 1988;165:209–215.

158 Fung MSC, Sun CRY, Gordon WL, Liou RS, Chang TW, Sun WNC, Daar ES, Ho DD: Identification and characterization of a neutralization site within the second variable region of human immunodeficiency virus type 1 gp120. J Virol 1992;66: 848–856.

159 Emini EA, Nara PL, Schleif WA, Lewis JA, Davide JP, Lee DR, Kessler J, Conley S, Matsushita S, Putney SD, Gerety RJ, Eichberg JW: Antibody-mediated in vitro neutralization of human immunodeficiency virus type 1 abolishes infectivity for chimpanzees. J Virol 1990;64:3674–3678.

160 Prince AM, Reesink H, Pascual D, Horowitz B, Hewlett I, Murthy KK, Cobb KE, Eichberg JW: Prevention of HIV infection by passive immunization with HIV immunoglobulin. AIDS Res Hum Retroviruses 1991;7:971–973.

161 Emini EA, Schleif WA, Nunberg JH, Conley AJ, Eda Y, Tokiyoshi S, Putney SD, Matsushita S, Cobb KE, Jett CM, Eichberg JW, Murthy KK: Prevention of HIV-1 infection in chimpanzees by gp120 V3 domain-specific monoclonal antibody. Nature 1992;355:728–730.

162 Berman PW, Gregory TJ, Riddle L, Nakamura GR, Champe MA, Porter JP, Wurm FM, Hershberg RD, Cobb EK, Eichberg JW: Protection of chimpanzees from infection by HIV-1 after vaccination with recombinant glycoprotein gp120 but not gp160. Nature 1990;345:622–625.

163 Girard M, Kieny M, Pinter A, Barré-Sinoussi F, Nara P, Kolbe H, Kusumi K, Chaput A, Reinhart T, Muchmore E, Ronco J, Kaczorek M, Gomard E, Gluckman J, Fultz PN: Immunization of chimpanzees confers protection against challenge with human immunodeficiency virus. Proc Natl Acad Sci USA 1991;88:542–546.

164 Goedert JJ, Mendez H, Drummond JE, Robertgu M, Minkhoff HL, Holman S, Stevens R, Rubinste A, Blattner WA, Willough A: Mother-to-infant transmission of HIV-1: Association with prematurity or low anti-gp120. Lancet 1989;ii:1351–1354.

165 Rossi P, Moschesse V, Broliden PA, Fundaró L, Quinti J, Plebani A, Giaquinto C, Tovo PA, Ljunggren K, Rosen J, Wigzell H, Jondal M, Wahren B: Presence of maternal antibodies to human immunodeficiency virus type 1 envelope glycoprotein gp120 epitopes correlates with the uninfected status of children born to seropositive mothers. Proc Natl Acad Sci USA 1989;86:8055–8058.

166 Parekh BS, Shaffer N, Pau C-P, Abrams E, Thomas P, Pollack H, Bamji M, Kaul A, Schochetman G, Rogers M, George JG: Lack of correlation between maternal antibodies to V3 loop peptides of gp120 and perinatal HIV-1 transmission. AIDS 1991; 5:1179–1184.
167 Halsey NA, Markham R, Wahren B, Boulos R, Rossi P, Wigzell H: Lack of association between maternal antibodies to V3 loop peptides and maternal-infant HIV-1 transmission. J AIDS 1992;5:153–157.
168 Geffin RB, Lai S, Hutto C, Scott GB, Scott WA, Master M, Parks WP: Quantitative analysis of human immunodeficiency virus type 1 antibody reactivity by Western immunoblots: Evaluation of relative antibody levels in seropositive individuals and mothers. J Infect Dis 1992;165:111–118.
169 O'Brien WA, Koyanagi Y, Namazie A, Zhoa J-Q, Diagne A, Idler K, Zack JA, Chen ISY: HIV-1 tropism for mononuclear phagocytes can be determined by regions of gp120 outside the CD4-binding domain. Nature 1990;348:69–73.
170 Westervelt P, Gendelman HE, Ratner L: Identification of a determinant within the human immunodeficiency virus type 1 surface envelope glycoprotein critical for productive infection of primary monocytes. Proc Natl Acad Sci USA 1991;88:3097–3101.
171 Hwang SS, Boyle TJ, Lyerly HK, Cullen BR: Identification of the envelope V3 loop as the primary determinant of cell tropism in HIV-1. Science 1991;253:71–74.
172 Takeuchi Y, Akutsu M, Murayama K, Shimizu N, Hoshino H: Host range mutant of human immunodeficiency virus type 1: Modification of cell tropism by a single point mutation at the neutralization epitope in the env gene. J Virol 1991;65:1710–1718.
173 Ivanoff LA, Looney DJ, McDanal C, Morris JF, Wong-Staal F, Langlois AJ, Petteway SR Jr, Matthews TJ: Alteration of HIV-1 infectivity and neutralization by a single amino acid replacement in the V3 loop domain. AIDS Res Hum Retroviruses 1991;7:595–603.
174 Chesebro B, Nishio J, Perryman S, Cann A, O'Brien W, Chen ISY, Wehrly K: Identification of human immunodeficiency virus envelope gene sequences influencing viral entry into CD4-positive HeLa cells, T-leukemia cells, and macrophages. J Virol 1991;65:5782–5789.
175 Cheng-Mayer C, Shioda T, Levy J: Host range, replicative, and cytopathic properties of human immunodeficiency virus type 1 are determined by very few amino acid changes in tat and gp120. J Virol 1991;65:6931–6941.
176 Freed EO, Myers DJ, Risser R: Identification of the principal neutralizing determinant of human immunodeficiency virus type 1 as a fusion domain. J Virol 1991; 65:190–194.
177 Freed EO, Risser R: Identification of conserved residues in the human immunodeficiency virus type 1 principal neutralizing determinant that are involved in fusion. AIDS Res Hum Retroviruses 1991;7:807–811.
178 Jong de JJ, Goudsmit J, Keulen W, Klaver B, Krone W, Tersmette M, de Ronde A: Human immunodeficiency virus type 1 clones chimeric for the envelope V3 domain differ in syncytium formation and replication capacity. J Virol 1992;66:757–765.
179 Fouchier RAM, Groenink M, Kootstra NA, Tersmette M, Huisman HG, Miedema F, Schuitemaker H: Phenotype associated sequence variation in the third variable domain of the HIV-1 gp120 molecule. J Virol 1992;66:3183–3187.
180 Callahan LN, Phelan M, Mallinson M, Norcross MA: Dextran sulfate blocks

antibody binding to the principal neutralizing domain of human immunodeficiency virus type 1 without interfering with gp120-CD4 interactions. J Virol 1991;65:1543–1550.

181 Ryu S-E, Kwong PD, Truneh A, Porter TG, Arthos J, Rosenberg M, Dai X, Xuong N, Axel R, Sweet RW, Hendrickson WA: Crystal structure of an HIV-binding recombinant fragment of human CD4. Nature 1990;348:419–426.

182 Camerini D, Seed B: A CD4 domain important for HIV-mediated syncytium formation lies outside the virus binding site. Cell 1990;60:747–754.

183 Truneh A, Buck D, Cassatt DR, Juszczak R, Kassis S, Ryu S-E, Healey D, Sweet R, Sattentau Q: A region in domain 1 of CD4 distinct from the primary gp120 binding site is involved in HIV infection and virus-mediated fusion. J Biol Chem 1991;266:5942–5948.

184 Autiero M, Abrescia P, Dettin M, di Bello C, Guardiola J: Binding to CD4 of synthetic peptides patterned on the principal neutralizing domain of the HIV-1 envelope protein. Virology 1991;185:820–828.

185 Hattori T, Koito A, Takatsuki K, Kido H, Katunuma N: Involvement of tryptase-related cellular protease(s) in human immunodeficiency virus type 1 infection. FEBS Lett 1989;248:48–52.

186 Koito A, Hattori T, Murakami T, Matsushita S, Maeda Y, Yamamoto T, Takatsuki K: A neutralizing epitope of human immunodeficiency virus type 1 has homologous amino acid sequences with the active site of inter-α-trypsin inhibitor. Int Immunol 1989;1:613–618.

187 Stephens PE, Clements G, Yarranton GT, Moore J: A chink in HIV's armour? Nature 1990;343:219.

188 Clements GJ, Price-Jones MJ, Stephens PE, Sutton C, Schulz TF, Clapham PR, McKeating JA, McClure MO, Thomson S, Marsh M, Kay J, Weiss RA, Moore JP: The V3 loops of the HIV-1 and HIV-2 surface glycoproteins contain proteolytic cleavage sites: A possible function in viral fusion? AIDS Res Hum Retroviruses 1991;7:3–16.

189 Wolfs TFW, Zwart G, Bakker M, Goudsmit J: HIV-1 genomic RNA diversification following sexual and parental virus transmission. Virology 1992;189:103–110.

190 Wolfs TFW, Zwart G, Bakker M, Valk M, Kuiken CL, Goudsmit J: Naturally occurring mutations within HIV-1 V3 genomic RNA lead to antigenic variation dependent on a single amino acid substitution. Virology 1991;185:195–205.

191 Wolfs TFW, de Jong J, van den Berg H, Tijnagel JMGH, Krone WJA, Goudsmit J: Evolution of sequences encoding the principal neutralization epitope of HIV-1 is host-dependent, rapid and continuous. Proc Natl Acad Sci USA 1990;87:9938–9942.

192 Köhler H, Goudsmit J, Nara P: Clonal antibody dominance: Cause for a limited and failing immune response to HIV-1 infection and vaccination. J AIDS, in press.

193 Nowak MA, Anderson RM, McLean AR, Wolfs TFW, Goudsmit J, May RM: Antigenic diversity thresholds and the development of AIDS. Science 1991;254:963–969.

194 Wahlberg J, Albert J, Lundeberg J, von Gegerfelt A, Broliden K, Utter G, Fenyö E-M, Uhlén M: Analysis of the V3 loop in neutralization-resistant human immunodeficiency virus type 1 variants by direct solid-phase DNA sequencing. AIDS Res Hum Retroviruses 1991;7:983–990.

195 Goudsmit J, Kuiken CL, Nara PL: Linear versus conformational variation of V3

neutralization domains of HIV-1 during experimental and natural infection. AIDS 1989;3(suppl 1):S119–S123.
196 Willey RL, Ross EK, Buckler-White AJ, Theodore TS, Martin MA: Functional interaction of constant and variable domains of human immunodeficiency virus type 1 gp120. J Virol 1989;63:3595–3600.
197 Ho DD, Moudgil T, Alam M: Quantitation of human immunodeficiency virus type 1 in the blood of infected persons. N Engl J Med 1989;321:1621–1625.
198 Coombs RW, Collier AC, Allain JP, Nikora B, Leuther M, Gjerset GF, Corey L: Plasma viremia in human immunodeficiency virus infection. N Engl J Med 1989; 321:1626–1631.
199 Koot M, Vos AHV, Keet RPM, de Goede REY, Dercksen MW, Terpstra FG, Coutinho RA, Miedema F, Tersmette M: HIV-1 biological phenotype in long-termed infected individuals evaluated with an MT-2 cocultivation assay. AIDS 1992;6:49–54.
200a Kuiken CL, de Jong JJ, Baan E, Keulen W, Tersmette M, Goudsmit J: Evolution of the V3 envelope domain in proviral sequences and isolates of human immunodeficiency virus type 1 during transition of virus biological phenotype. J Virol 1992;66: 4622–4627.
200b De Jong JJ, De Ronde A, Keulen W, Tersmette T, Goudsmit J: Minimal requirements for the human immunodeficiency virus type 1 V3 domain to support the syncytium-inducing phenotype: analysis by single amino acid substitution. J Virol, in press.
201 Roos MThL, Lange JMA, de Goede REY, Coutinho RA, Schellekens PThA, Miedema F, Tersmette M: Virus phenotype and immune response in primary HIV-1 infection. J Infect Dis, in press.
202 Myers G, Korber B, Berzofsky JA, Smith RF, Pavlakis GN: Human retroviruses and AIDS. Theor Biol Biophys, Los Alamos, 1991.
203 Javaherian K, Langlois AJ, LaRosa GJ, Profy AT, Bolognesi DP, Herlihy WC, Putney SD, Matthews TJ: Broadly neutralizing antibodies elicited by the hypervariable neutralizing determinant of HIV-1. Science 1990;250:1590–1593.
204 Ohno T, Terada M, Yoneda Y, Shea KW, Chambers RF, Stroko DM, Nakamura M, Kufe DW: A broadly neutralizing monoclonal antibody that recognizes the V3 region of human immunodeficiency virus type I glycoprotein gp120. Proc Natl Acad Sci USA 1991;88:10726–10729.
205 Holley LH, Goudsmit J, Karplus M: Prediction of optimal peptide mixtures to induce broadly neutralizing antibodies to human immunodeficiency virus type 1. Proc Natl Acad Sci USA 1991;88:6800–6804.
206 Korber B, Wolinski S, Haynes B, Kunstman K, Levy R, Furtado M, Myers G: HIV-1 intrapatient sequence diversity in the immunogenic V3 region. AIDS Res Hum Retroviruses, in press.

Prof. Dr. J. Goudsmit, Department of Virology, University of Amsterdam,
Academic Medical Centre, Meibergdreef 15,
NL–1105 AZ Amsterdam (The Netherlands)

Norrby E (ed): Immunochemistry of AIDS.
Chem Immunol. Basel, Karger, 1993, vol 56, pp 34–60

B Cell Antigenic Site Mapping of HIV-1 Glycoproteins[1]

A. Robert Neurath

Laboratory of Biochemical Virology, Lindsley F. Kimball Research Institute of the New York Blood Center, New York, N.Y., USA

There are two envelope (env) glycoproteins expressed on the surface of the human immunodeficiency virus type 1 (HIV-1), the surface glycoprotein gp120 and the transmembrane glycoprotein gp41 [1]. These two glycoproteins are gene products of a single *env* gene and are synthesized as a precursor glycoprotein gp160 which is proteolytically cleaved into gp120 and gp41. There are several hundred copies of each env glycoprotein per virion arranged symmetrically on the virus surface, probably in the form of noncovalently linked trimers consisting of gp120/gp41 pairs, the two distinct glycoproteins being associated noncovalently (fig. 1). The gp120 is spontaneously shed from both virus particles and infected cells. This shedding is enhanced by the association of gp120 with the CD4 molecule, the cellular receptor for HIV-1 [9]. The propensity to shed gp120 from HIV-1 virions is different for distinct HIV-1 isolates, probably due to differences in the strength of gp120-gp41 association for distinct HIV-1 isolates, influenced by amino acid sequence variability (fig. 2) within regions of both gp120 and gp41 involved in intersubunit interaction (fig. 1). This is supported by the finding that the association constant for the gp120-CD4 interaction is essentially the same for gp120 glycoproteins corresponding to distinct HIV-1 isolates greatly differing from each other with respect to virus neutralization by soluble CD4 [12–15].

Both gp120 and the external portion of gp41 are glycosylated (fig. 3, 4). There are five potential N-glycosylation sites on gp41 from HIV-1-IIIB. Four of these sites are mostly conserved among distinct HIV-1 isolates [11].

[1] This study was supported by grants CA-43315 and AI-29373 from the National Institutes of Health (USA). We thank R. Ratner for computer graphics and preparation of the manuscript and R. Morgan and L. Morgan for drawings.

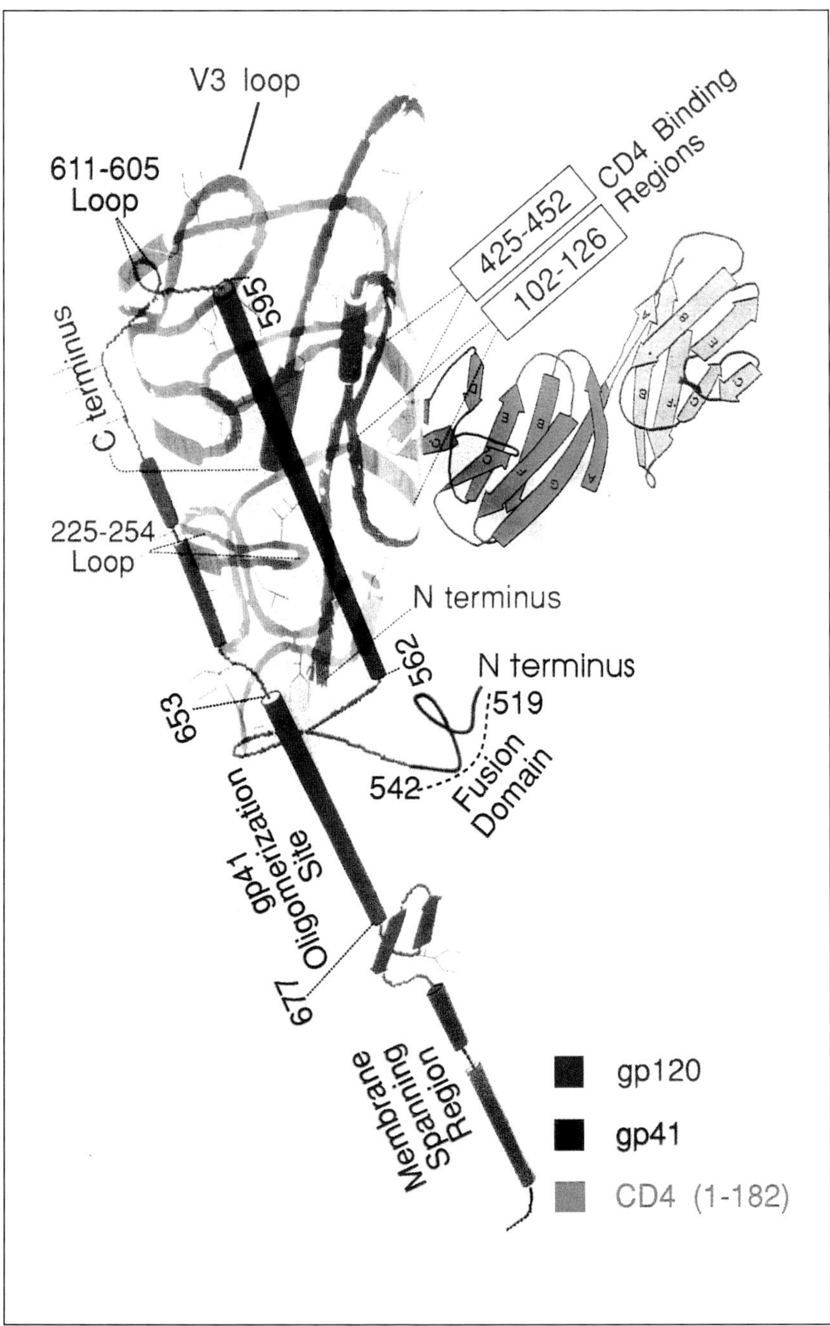

Fig. 1. (For legend see reverse side.)

Fig. 1. Schematic representation of an HIV-1 gp120/gp41 monomer shown in association with the N-terminal 182 amino acid residues of CD4 attached to the monomer. The gp120/gp41 monomers associate into polymeric structures (probably trimers), representing spikes on the virus envelope [2]. The portion of gp41 spanning and extruding from the lipid bilayer is shown in blue and represents a modification of the structure proposed by Gallaher et al. [3]. The glycoprotein gp120, shown in red, is a folded version of a proposed schematic structure [4]. The proposed attachment sites for gp41 on gp120 are located on the N-terminal and C-terminal segments of gp120, and on portions of the V3 (303-338) loop and a loop encompassing residues (224-254). The binding site for gp120 on gp41 probably encompasses a 'leucine zipper'-like repeat sequence (566-594). The (637-666) segment of gp41 probably includes residues involved in gp41 oligomerization. The N-terminus of gp41 [residues (519-542)] encompasses the domain of this glycoprotein mediating fusion of HIV-1 virions with cells. The latter segment of gp41 probably becomes available on the surface of virions after they have attached to the CD4 receptor. The V3 disulfide loop on gp120 and the loop between cysteine residues (605-611) are likely to be in proximity [4]. Both disulfide bonds are essential for virus infectivity and the post-translational processing of gp160 into gp120 and gp41 [5–7]. The cylindric structures within gp120 and gp41 represent α-helices. The position of saccharide chains in gp120/gp41 is indicated by symbols —< or ⊤⊂. The N-terminal half of the extracellular portion of CD4 is shown in green, and is in contact with gp 120 through residues CD4 (38-52) including the C″ ridge of the CD4 domain 1 (shown in darker green) [8]. Nonadjacent residues in gp120, encompassing residues (102-126) and (391-452) appear to contribute to the binding site for CD4.

Fig. 3. Antigenicity of peptides from gp120 as determined by their reactivity with antibodies from sera of HIV-1-infected individuals. The mean endpoint dilutions of sera from 18 infected individuals were determined by ELISA using polystyrene plates coated with the respective peptides [16]. The range of endpoint titers of antibodies reacting with distinct segments of gp120 is represented by a color scheme superimposed on the proposed [4] unfolded structure of gp120. Glycosylation sites containing high mannose-type and/or hybrid-type oligosaccharide structures are indicated by (—<) and glycosylation sites containing complex-type oligosaccharide structures are indicated by (⊤⊂) [17]. Numbers in italics indicate selected positions of cysteine residues. Other numbers indicate N- and C-termini and glycosylation sites. The peptides were from the sequence of HIV-1 IIIB [18]. However, antibodies recognizing the V3 loop (303-338) were determined by ELISA using wells coated with a mixture of peptides corresponding to full-length V3 loops from 21 distinct HIV-1 isolates (clones) [19]. Larger numbers in italics indicate positions of cysteine residues involved in disulfide bonds [17]. Smaller numbers indicate sites of attachment of oligosaccharide chains [17]. Red lines on the periphery of gp120 indicate stretches of 5 or more conserved amino acids. The second red line from the N-terminus corresponds to the (56-63) segment.

Fig. 4. Antigenicity of peptides from gp41 as determined by their reactivity with antibodies from sera of HIV-1-infected individuals. The range of endpoint titers of antibodies reacting with distinct segments of gp41 is represented by a color scheme superimposed on the proposed structure of gp41 [3, 4]. Potential glycosylation sites are indicated by (⊤⊂). Position of the (605-611) disulfide loop, of glycosylation sites and of other selected segments are indicated. For additional explanations, see legend to figure 3.

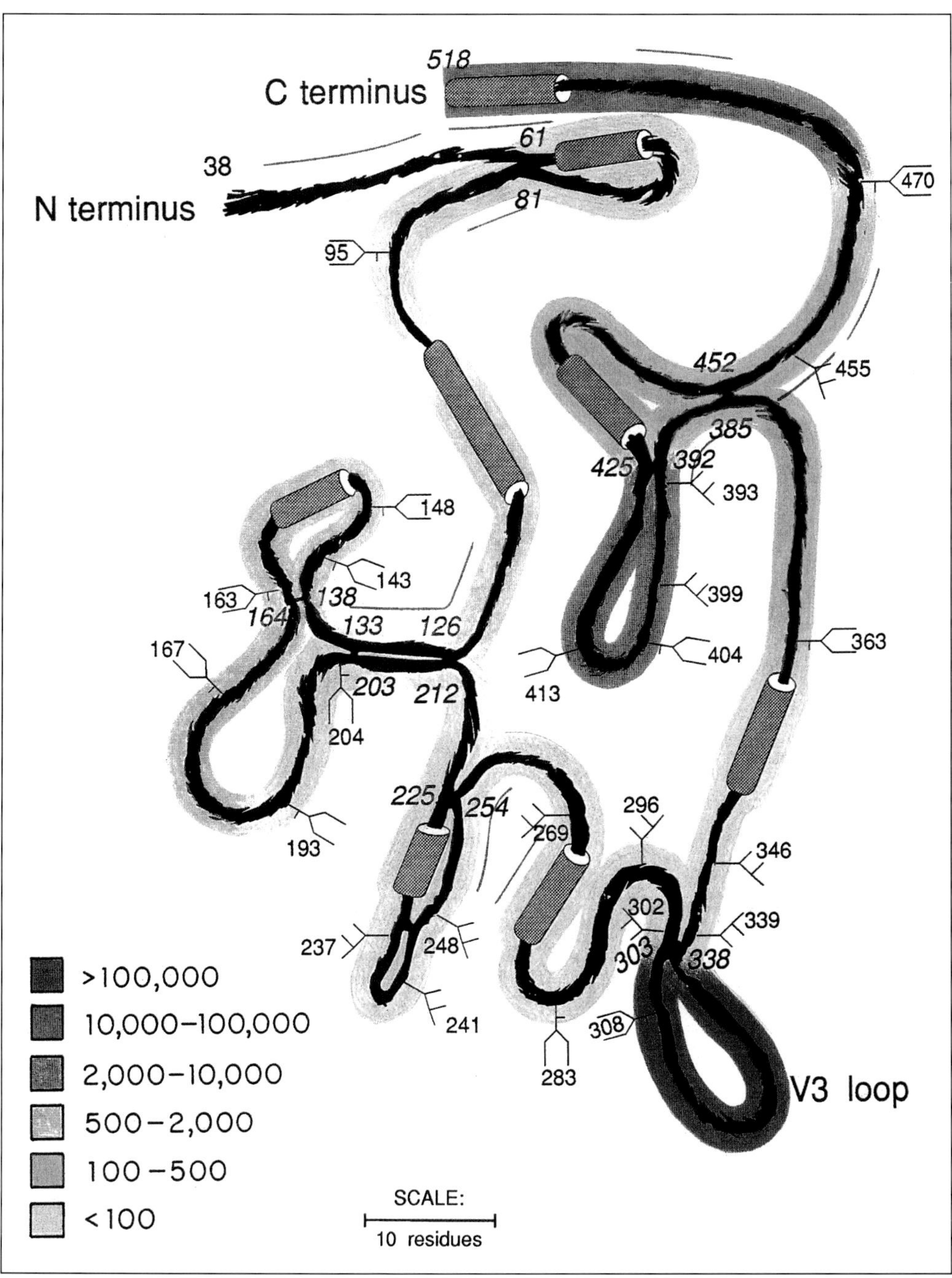

Fig. 3. (For legend see opposite page.)

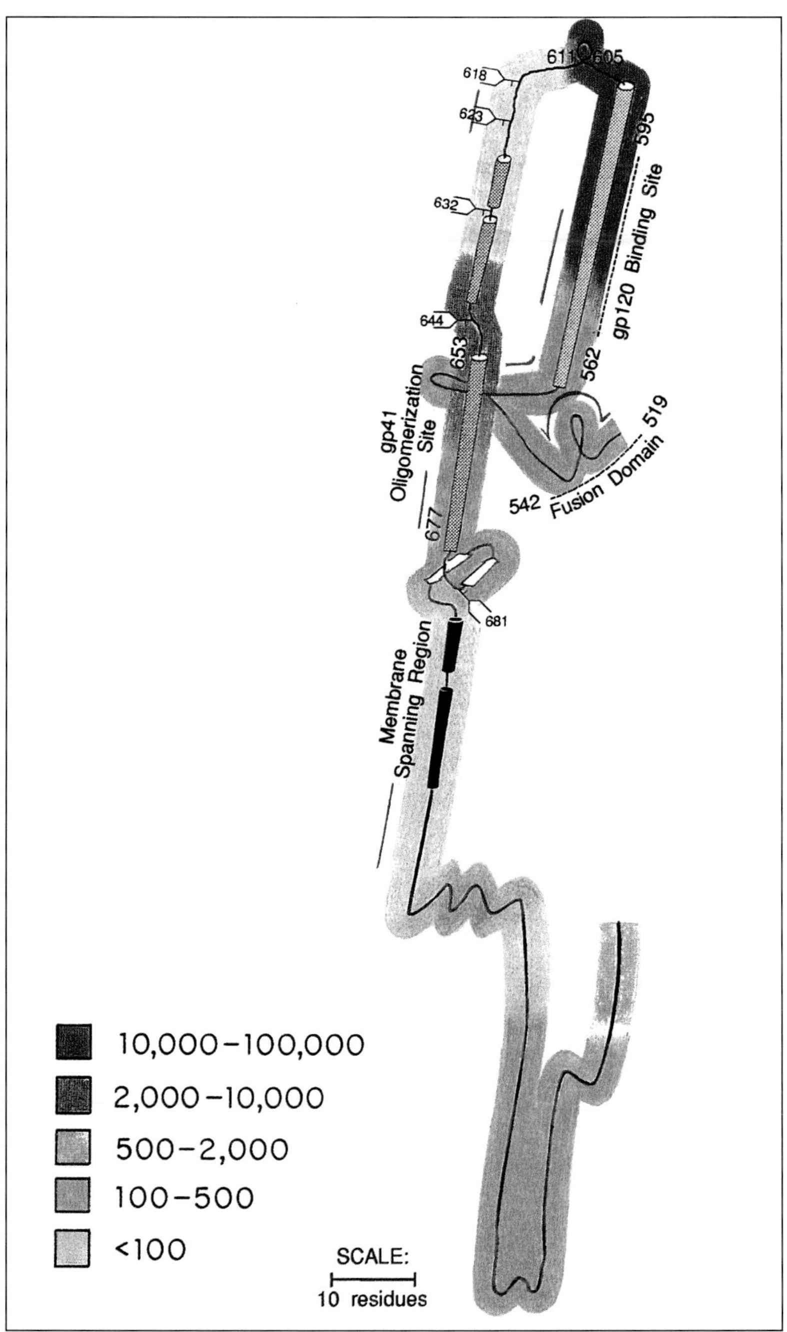

Fig. 4. (For legend see page before fig. 3.)

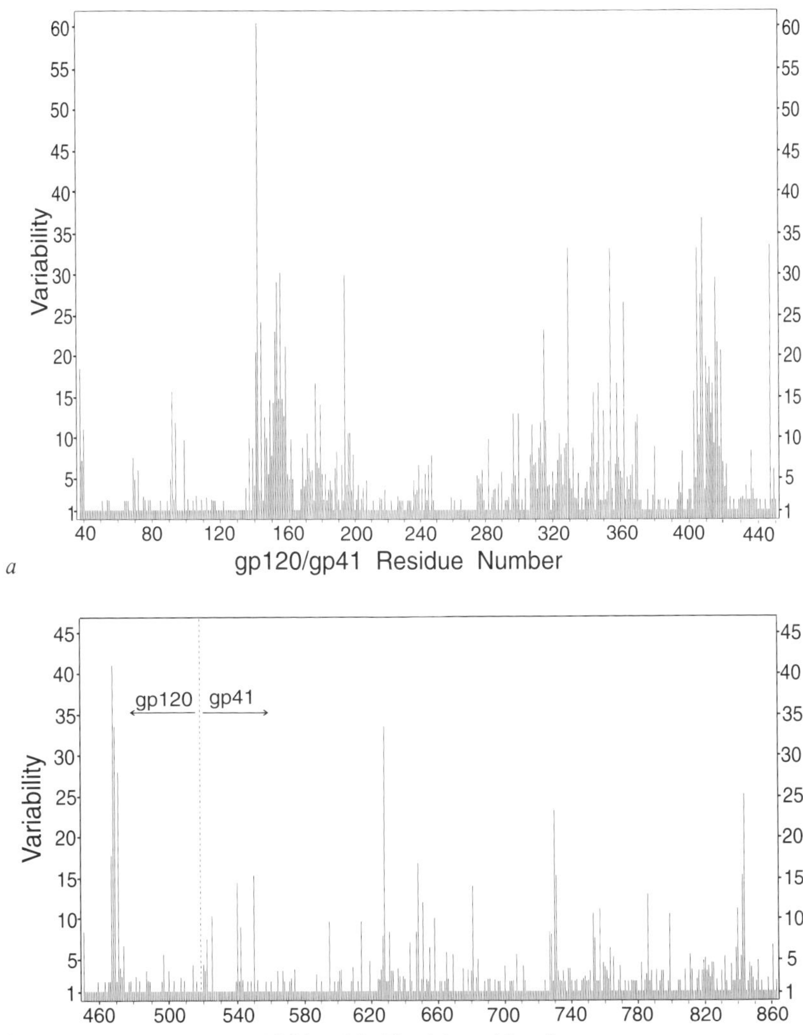

Fig. 2. Variability of the amino acid sequence of the HIV-1 glycoproteins gp120 and gp41. Variability analysis was done according to Wu and Kabat [10]. The variability was calculated from 30 distinct HIV-1 gp120/gp41 sequences [11] according to the formula:

$$\text{variability} = \frac{\text{number of different amino acids at a given position}}{\text{frequency of the most common amino acid at that position}} \times n,$$

where n = the number of sequences analyzed. The first amino acid plotted on the abscissa is number 38, corresponding to the N terminus of mature gp120. The C terminus of gp120 and the N terminus of gp41 correspond to residues 518 and 519, respectively. The entire gp120/gp41 sequence was arbitrarily divided for clarity into two portions *(a, b)*.

Amino acid substitutions introduced by in vitro mutagenesis resulting in selective elimination of N-glycosylation sites may impair viral infectivity, suggesting that N-linked sugars may have an important role in HIV-1 replication [20, 21]. About half of the mass of gp120 is contributed by carbohydrate. There are 24 N-glycosylation sites on gp120 from HIV-1-IIIB [17], but the number and position of N-glycosylation sites may differ in different HIV-1 isolates [11]. Elimination of N-glycosylation sites around the CD4-binding site by mutation resulted in decreased infectivity [22]. Antibodies against galactosyl ceramide and mucin-type carbohydrate neoantigens were reported to neutralize the infectivity of HIV-1 [23, 24]. Notwithstanding these findings, antibodies specific for the protein moiety of HIV-1 env glycoproteins seem to be of primary biological importance, and are the subject of this review.

There are several compelling reasons for defining antibody-binding sites (B cell epitopes) on gp120/gp41: (1) design of diagnostic reagents for detecting anti-HIV-1 antibodies and for serotyping HIV-1 subtypes; (2) development of prophylactic and/or therapeutic measures based on active or passive immunization; (3) development of antiviral drugs having a predetermined target site [25, 26]; (4) definition of functionally important sites playing a role in virus assembly and in virus-cell interactions [4]; (5) recognition of sites potentially eliciting autoimmune responses (including but not restricted to sites on cellular proteins having partial primary sequence homology with gp120/gp41 [27–35], possibly leading to programmed cell death (apoptosis) [36–38]; (6) characterization of antibodies causing enhancement of HIV-1 infection [39]; (7) identification of sites on gp120/gp41 playing a role in tropism of HIV-1 for cells of distinct lineages (T4+ lymphocytes, monocytes, etc.) [40–46], potentially permitting the serological distinction between HIV-1 isolates with different cell tropisms, and (8) definition of the effect of glycosylation on antigenicity of gp120/gp41 [47].

As is the case with viral proteins in general, there are continuous and discontinuous [48] B cell epitopes on gp120/gp41. Most protein epitopes are discontinuous, and cross-reacting peptides may mimic a portion of an epitope [49]. When characterizing B cell epitopes on viral proteins, it is necessary to distinguish from each other epitopes: (1) which are exposed on individual glycoprotein subunits but become inaccessible due to oligomerization of these subunits within the viral envelope (= cryptotopes); (2) which result from assembly of the viral envelope but are absent within the constituent monomeric units (= neotopes); (3) which are exposed on both the subunits and the assembled viral envelope (= metatopes) [48] (fig. 1), and

Table 1. Methods to localize B cell epitopes on viruses and viral proteins

Method	Type of epitope recognized
1. X-ray crystallography of antigen-antibody (Fab) complexes	Mostly discontinuous epitopes reacting with homologous antibody
2. Study of cross-reactive binding of natural or synthetic peptide fragments with antiviral antibodies	Continuous epitopes cross-reacting with antiviral antibody
3. Study of cross-reactive binding of recombinant fusion proteins with anti-viral antibodies	Cross-reactive, mostly continuous epitopes
4. Neutralization by antiviral monoclonal antibodies (MoAbs) of filamentous bacteriophages expressing on their surface fragments of viral proteins	Cross-reactive, mostly continuous epitopes
5. Study of cross-reactive binding of virus or viral proteins with anti-peptide antibodies	Cross-reactive continuous epitopes
6. Proteolysis of virus protein-MoAb complexes followed by separation and partial sequencing of the cleavage fragments	Continuous and discontinuous epitopes
7. Analysis of viral mutants with MoAbs	Neutralization epitopes and discontinuous epitopes
8. Competitive binding assays with pairs of MoAbs or of MoAb and antipeptide antibodies	Only relative position of epitopes
9. Identification of critical residues in peptide fragments by systematic replacement or deletion studies	Continuous epitopes containing essential residues interspersed with residues irrelevant for antibody binding
10. Site-directed mutagenesis of viral proteins followed by studies of MoAb binding to the mutant proteins	Continuous, discontinuous and neutralization epitopes

(4) neutralization epitopes (metatopes or neotopes) which are recognized by antibody molecules neutralizing virus infectivity. The accessibility of distinct epitopes on the virion surface may also be altered by its attachment to cell receptors, in the case of HIV-1, the CD4 molecule, and possibly other cell membrane proteins.

Methods to localize B cell epitopes on viruses and viral proteins [4, 48, 50–54] are summarized in table 1. Results described in this review were mostly obtained by methods 2, 5, 8 and 9.

Amino Acid Sequence Variability of HIV-1 gp120/gp41

HIV-1 undergoes sequence variations during in vivo and in vitro replication. Therefore, it was proposed that HIV-1 isolates cannot be described in defined molecular terms as a single species and should be considered only as quasispecies [55, 56]. The sequence variability is the highest for the env glycoprotein gp120 and to a lesser degree for gp41 (fig. 2). The sequence homology among gp120 glycoproteins from distinct HIV-1 isolates is ~40% [11] and there are only ten short (5- to 13-amino-acid-residue) stretches in gp120 which are completely conserved, representing only ~15% of the entire gp120 sequence (fig. 3). The homology between gp41 sequences corresponding to distinct HIV-1 isolates is ~52%, and there are only six completely conserved stretches 5–13 amino acids long, corresponding to 15% of the entire gp41 sequence (fig. 4). On the other hand, cysteine residues involved in formation of disulfide bridges are conserved both in gp120 and gp41, suggesting that the basic architecture of these two glycoproteins is maintained among the distinct HIV-1 isolates. However, conformational changes elicited by distal amino acid substitutions may alter the immunological specificity and/or surface accessibility of biologically important immunodominant epitopes, the primary amino sequence of which remains constant in virus variants [57].

Considering the above findings, the results of B cell epitope mapping obtained with one HIV-1 isolate (clone) cannot be automatically extrapolated to other or all HIV-1 isolates. In this respect, it is especially unfortunate that sequences corresponding to historically first identified HIV-1 isolates [18, 58], having a low prevalence (< 5%, as calculated by LaRosa et al. [59]) in the worldwide population of infected individuals, have become the predominant basis for development of antisera, recombinant proteins and synthetic peptides for mapping B and T cell epitopes on HIV-1 proteins. Results of B cell epitope mapping of gp120/gp41, presented in the subsequent portions of this review, will have to be evaluated with these reservations in mind.

Mapping of B Cell Epitopes on gp120/gp41 Using Peptides

Due to efforts to develop diagnostic reagents for detection of anti-HIV-1 antibodies and protective immunogens against HIV-1 infection, attempts to define B cell epitopes on HIV-1 glycoproteins were initiated immediately after the first data on the nucleotide sequence of the viral genome became available [18, 58, 60] and have been continuing since then [for review, see ref. 16, 19 and 61–63]. Most of these studies were carried out using method 2 (table 1) with synthetic peptides derived from the HIV-1-IIIB LAV sequences [11, 58] and with sera from HIV-1-infected humans. The occurrence of a multitude of HIV-1 variants in infected individuals and the predominance of HIV-1-MN- and -SC-like viruses, documented by both direct nucleotide sequencing and serological methods [59, 64–67], has not been fully appreciated in efforts to define B cell epitopes on HIV-1 glycoproteins except for studies involving the V3 hypervariable loop of gp120 (see below). We utilized relatively long (19- to 36-residue) synthetic peptides to map B cell epitopes in order to increase the chance that the peptides may also mimic discontinuous epitopes and disulfide loops within the env glycoproteins [16, 19, 61, 63]. Instead of merely determining whether or not the synthetic peptides from gp120/gp41 react with human anti-HIV-1, the dilution endpoints of the antisera with each of the peptides were determined, allowing the hierarchical arrangement of segments of gp120/gp41 according to their reactivity with human anti-HIV-1. It was recognized in the course of these studies that individual human anti-HIV-1 sera differed from each other with respect to recognition of the distinct peptides [16]. The antibody titers were averaged in order to generate B cell epitope maps (fig. 3, 4). Most peptides from gp120 (fig. 3) and gp41 (fig. 4) were recognized in general by human anti-HIV-1, although sera from individual patients frequently failed to recognize some of the peptides. However, the dilution endpoint titers for distinct peptides differed by $> 10^4$. The immunodominant epitope of gp120 was localized within the V3 loop (residues 303–338) followed by the C-terminal portion of gp120 (fig. 3). The immunodominant epitope of gp41 was localized within the segment 579–611 which includes the only disulfide loop in gp41 (residues 605–611), followed by an epitope within the 637–666 segment (fig. 4). The latter segment is inaccessible to antibodies within gp41 oligomers [4] (fig. 1). The sera used for B cell epitope mapping were also titered for antibodies recognizing recombinant gp120 and gp160, respectively. Results in table 2 show that the mean endpoint titer determined by immunoassays with gp120 was by one order of magnitude lower than the combined titers measured with peptides from gp120 which included peptides from the V3 loop of 21 distinct HIV-1 isolates, indicating that gp120 from HIV-1-IIIB should not be considered in general as a reagent of choice for detection of anti-gp120 antibodies in sera from HIV-1-infected individuals. The mean endpoint titer determined by

Table 2. Mean antibody titers of sera from 18 HIV-1-infected individuals measured by immunoasssays with recombinant gp120-IIIB and gp160-IIIB, respectively, as compared with titers determined by immunoassays with synthetic peptides from gp120/gp41

Dilution endpoints determined with	Reciprocal of dilution endpoint
gp120-IIIB	17,000
gp120 peptides[1]	180,000[2]
gp160-IIIB	870,000
gp120[1]+gp41 peptides	257,000[2]
gp41 peptides	76,000[2]

[1] Peptides from the gp120-IIIB sequence + a mixture of peptides from 21 distinct HIV isolates [73] were used to detect antibodies to the V3 loop.
[2] The numbers correspond to the sum of reciprocal dilution endpoints determined with each peptide.

immunoassays with recombinant gp160 (= gp120+gp41) was about 50-fold higher in comparison with the anti-gp120 titer and about 13-fold higher than that determined with the synthetic peptide 579–611, suggesting that (1) either most of the antibodies in human sera are directed against gp41 rather than gp120 or that this predominance is the result of greater sequence variability within gp120 as compared with gp41 (fig. 2), resulting in greater mismatch of anti-gp120 as compared with anti-gp41-specific antibodies in human sera with HIV-1-IIIB-derived recombinant glycoproteins, and (2) the synthetic peptides from gp41 only partially mimic epitopes located on the gp41 glycoprotein.

Since the anti-gp120/gp41 antibodies in human sera were mostly not elicited by HIV-1-IIIB, but by other HIV-1 isolates having only partial sequence homology with HIV-1-IIIB, the results of B cell epitope mapping shown in figures 3 and 4 may not accurately reflect the hierarchy of immunodominance of distinct regions in gp120/gp41. For this reason, the B cell epitope mapping studies were repeated with antisera raised against recombinant gp120-IIIB and gp160-IIIB, respectively. (The rabbit antisera were obtained from American BioTechnologies, Cambridge, Mass., and MicroGeneSys, West Haven, Conn., respectively.) The immunodominant epitope discerned by the anti-gp120 antibodies corresponded again to the V3 loop (fig. 5), but the hierarchy of immunodominance of distinct regions in gp120 was not the same as the one arrived at by using human anti-HIV-1.

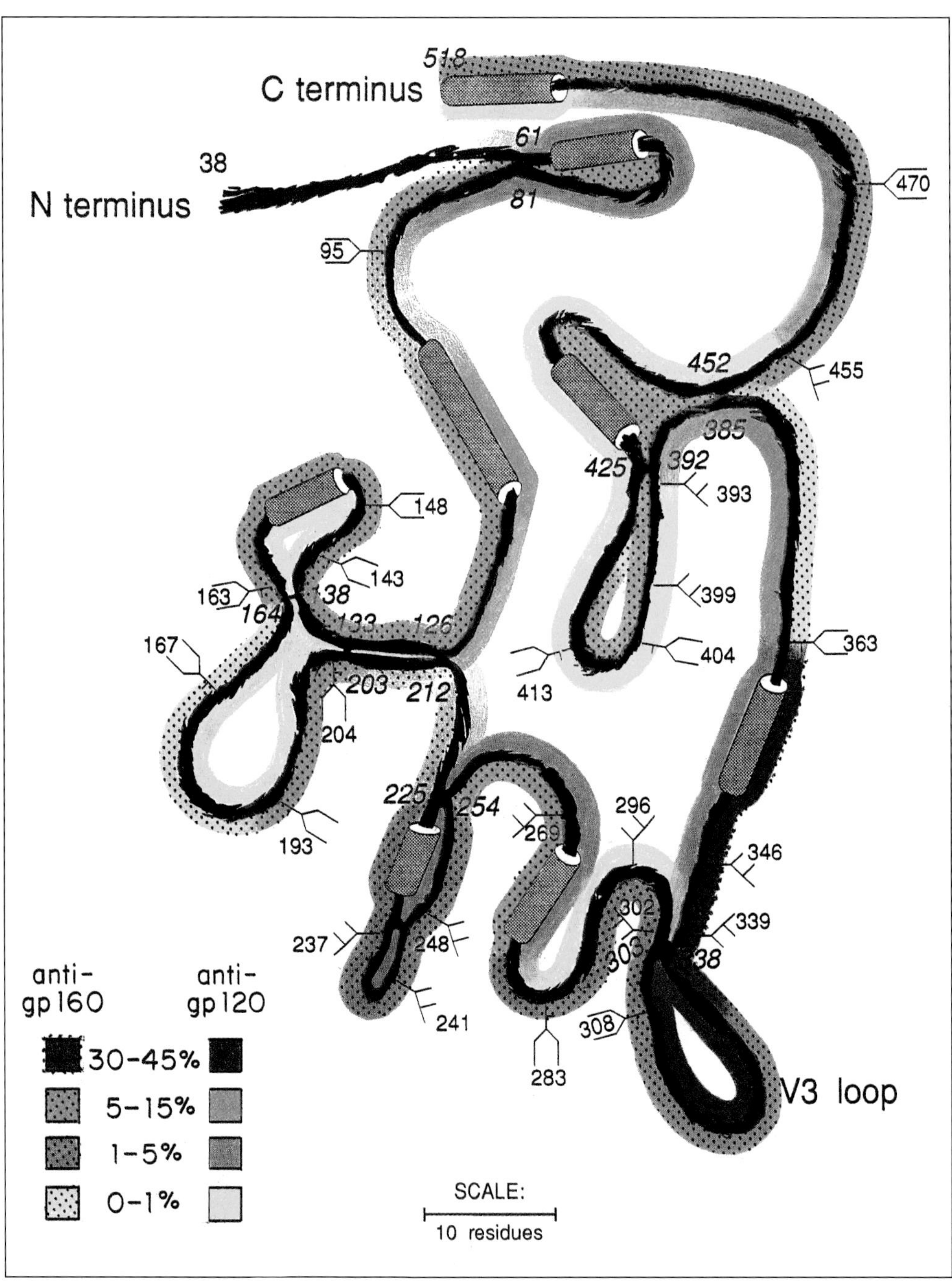

Fig. 5. (For legend see reverse side.)

Fig. 5. Recognition of peptides from gp120 by rabbit antisera raised against gp120 and uncleaved gp160, respectively, both from HIV-1 IIIB [18]. Antibody titers of the two antisera were determined using peptides from gp120. Relative endpoint dilutions for each peptide were determined from the formula:

$$\frac{\text{Endpoint dilution for a given peptide}}{\text{Sum of endpoint dilution values for all peptides}} \times 100$$

The calculated percentages, graded according to the color scheme shown in the insert, were superimposed on segments of the proposed folded schematic structure of gp120. For additional explanations, see legend to figure 3.

Fig. 6. Antigenic similarity between peptides from gp120 and the corresponding sequences of gp120. The elicitation by synthetic peptides of antibodies reacting with the homologous peptide and with recombinant gp160 (= uncleaved gp120 + gp41) was determined [61]. The titers of antibodies against gp160 and the homologous peptide were compared and expressed in percentages based on the following formula:

$$\text{Antigenic similarity (\%)} = \frac{\text{endpoint titer of antibodies reacting with gp160}}{\text{endpoint titer of antibodies reacting with peptide}} \times 100$$

The calculated results, graded according to a color scheme shown in the insert, were superimposed on segments of the schematic unfolded structure proposed for gp120 [4]. A calculated antigenic similarity value of 100% does not imply that the epitopes on the glycoprotein and the synthetic peptide are identical. For additional explanations, see legend to figure 3.

Fig. 7. Antigenic similarity between peptides from gp41 and the corresponding sequences of gp41. The elicitation by synthetic peptides of antibodies reacting with the homologous peptide and with recombinant gp160 (= uncleaved gp120 + gp41) was determined [61]. The titers of antibodies against gp160 and the homologous peptide were compared and expressed in percentages based on the following formula:

$$\text{Antigenic similarity (\%)} = \frac{\text{endpoint titer of antibodies reacting with gp160}}{\text{endpoint titer of antibodies reacting with peptide}} \times 100$$

The calculated results, graded according to a color scheme shown in the insert, were superimposed on segments of the schematic unfolded structure proposed for gp41 [3, 4, 7]. For additional explanations, see legends to figures 4 and 6.

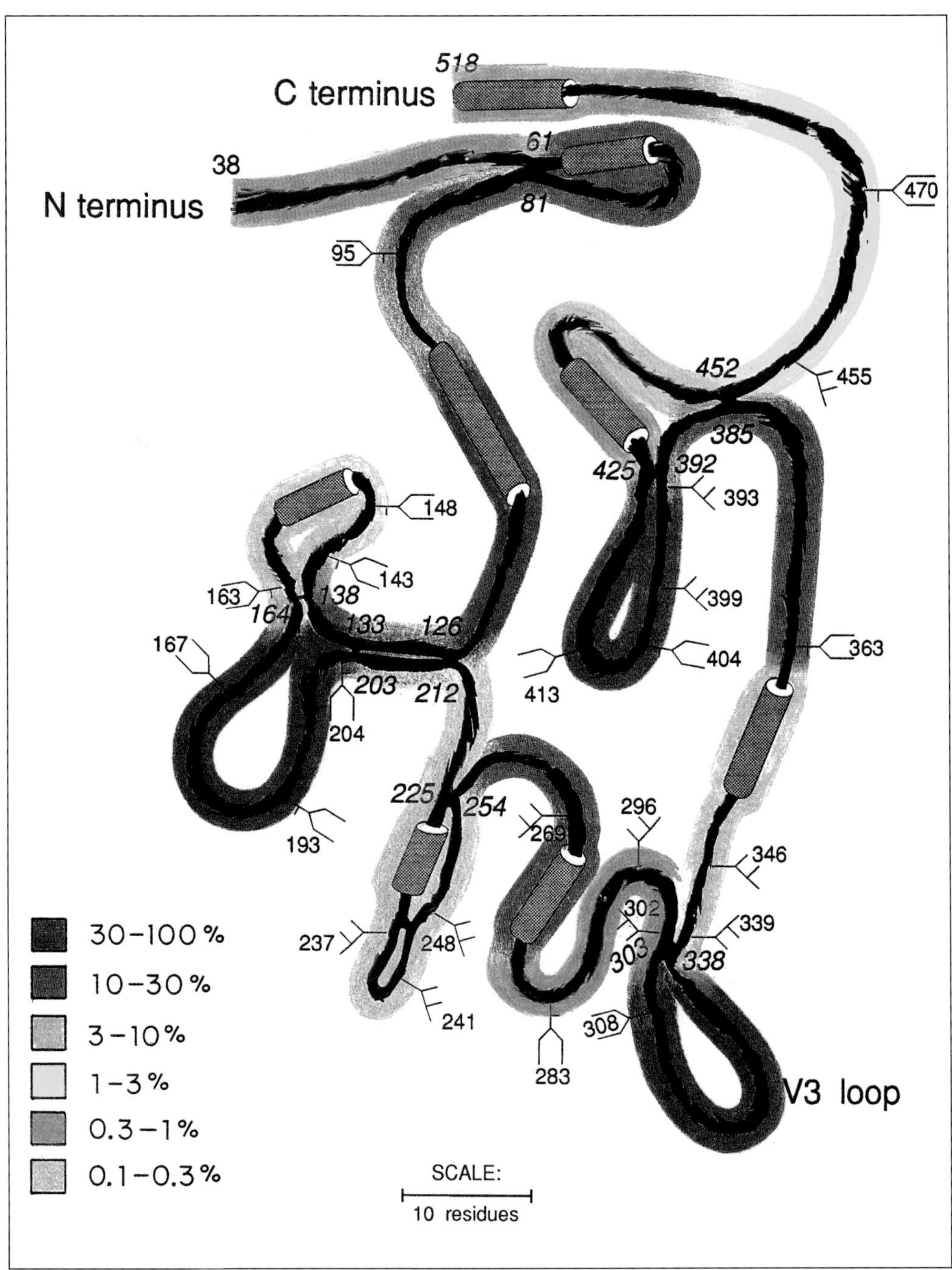

Fig. 6. (For legend see opposite page.)

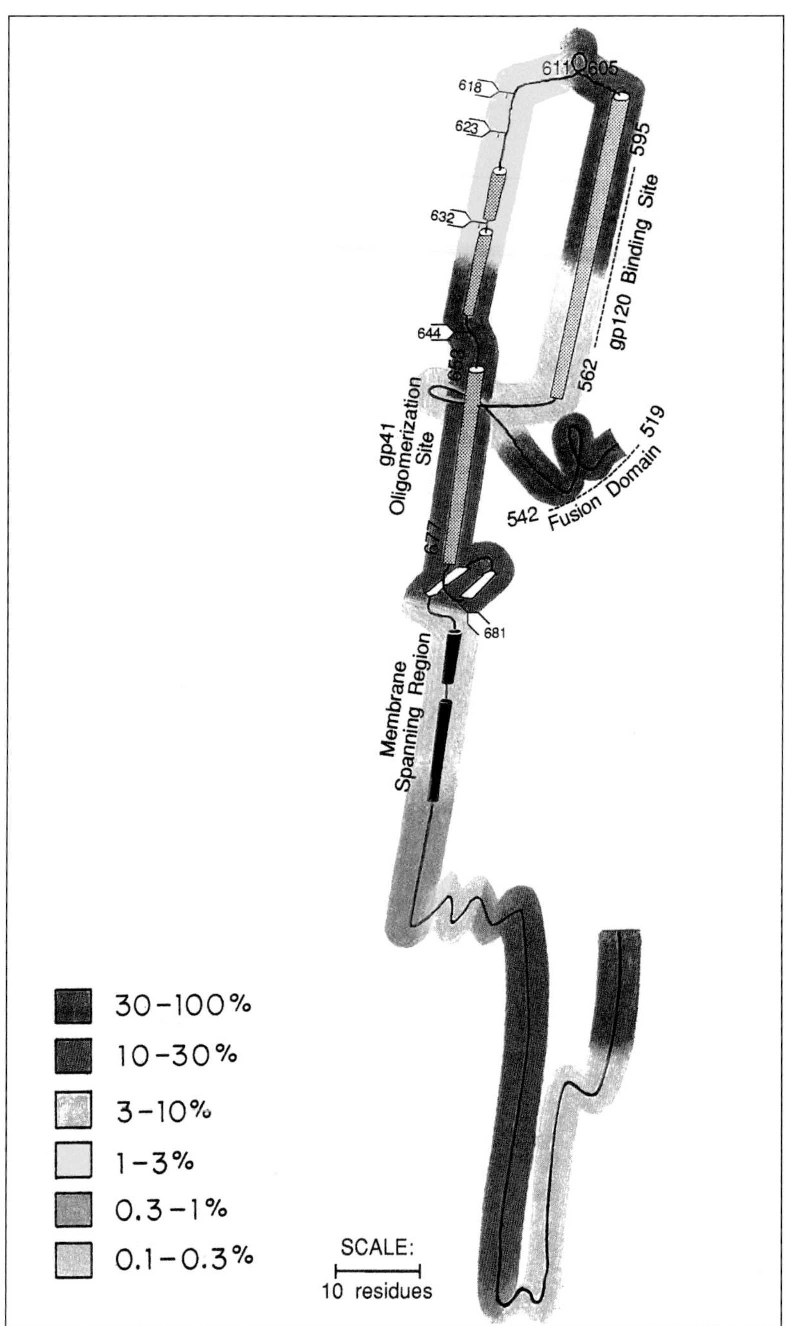

Fig. 7. (For legend see page before fig. 6.)

Since antibodies to regions other than the V3 loop appear to be more prevalent in the rabbit antisera in comparison with human anti-HIV-1, this observation may be the result of complete sequence homology between synthetic peptides and the corresponding segments on gp120 used for immunization or may be the reflection of species-specific differences in the repertoire of antibodies elicited by the same antigen in different species [68]. Surprisingly, the hierarchy of immunodominance obtained with anti-gp160 was different from that established with anti-gp120 (fig. 5). Most significantly, the V3 loop did not appear to be the immunodominant epitope when anti-gp160 was used for mapping. Rather, the region 331–361, partly overlapping and adjoining the V3 loop, was recognized as immunodominant. The different patterns of immunodominance observed with anti-gp120 as compared with anti-gp160 may possibly explain the protection of chimpanzees from HIV-1 infection after immunization with recombinant gp120 but not with gp160 [69].

The only peptides from gp41 recognized by rabbit antibodies to gp160-IIIB were 579–611, 637–666 and 729–758 (antibody dilution endpoints 1/28,000, 1/20,000 and 1/5,600, respectively), indicating that the 579–611 and 637–666 segments of gp41 were discerned as immunodominant epitopes by both human anti-HIV-1 and rabbit anti-gp160-IIIB.

gp120/gp41 B Cell Epitope Mapping Using Antipeptide Antisera

The availability of antisera to peptides from gp120/gp41 [4, 19, 61, 63] and of recombinant gp120 and gp160 offered another opportunity for B cell epitope mapping (method 5, table 1) without limitations imposed by partial sequence homology between the subtype of antigens and antibodies used for mapping (fig. 3, 4). The distinct segments of gp120/gp41, corresponding to peptides used as immunogens, were ranked according to the extent of their recognition by the anti-peptide antibodies (fig. 6, 7). The ranking of B cell epitopes by this method is independent of the immunogenicity of gp120/gp41 segments in distinct species, and reflects the extent of antigenic similarity between synthetic peptides and the corresponding segments in gp120/gp41. The peptides antigenically most similar to the corresponding gp120/gp41 segments are: 303–338 (= the V3 loop), 102–126, 361–392, 391–425 and a loop encompassing residues 133–138 and 164–203, containing four cysteine residues (fig. 6). Synthetic peptides corresponding to segments of the latter loop are more dissimilar to the corresponding gp120 segments in comparison with the full-length loop peptide (data not shown). The antigenic similarity between the V3 loop of gp120 and the corresponding synthetic peptide was confirmed also for gp120 from HIV-1-SF2, the only commercially available gp120 other than that from HIV-1-IIIB. Synthetic peptides with the greatest antigenic similarity to the corresponding segments of gp41 were: 579–611, 637–666, 518–542 and 845–862 (fig. 7). The latter two peptides are from the N and C termini of gp41.

Synthetic Peptides as Diagnostic Reagents to Detect and Subtype Anti-HIV-1 Antibodies

B cell epitope mapping studies revealed that peptides encompassing the 605–611 disulfide loop of gp41 and those from the V3 loop of gp120 are optimally recognized by sera from HIV-1-infected individuals. Thus, peptides from these two regions of gp120/gp41 are the best candidates for the design of simple site-specific serological tests for detecting immune responses to HIV-1. However, neither of these two regions of HIV-1 glycoproteins has conserved amino acid sequences (fig. 2). Since the immunodominant B cell epitope from gp41 has a more conserved amino acid sequence in comparison with the V3 loop (fig. 2, 8, 9), the corresponding peptide from gp41 is a preferable reagent for demonstration of anti-HIV-1 antibodies in human sera. In addition, the level of antibodies specific for the V3 loop appears to decrease more rapidly during the course of development of the acquired immunodeficiency syndrome, as compared with the level of antibodies specific for the dominant B cell epitope of gp41 [16]. Therefore, peptides mimicking the latter epitope(s) represent reagents of preferred choice for site-directed serology for detection of anti-HIV-1 antibodies. Peptides from both the immunodominant B cell epitope regions of gp41 and gp120 are also potential reagents for subtyping of anti-HIV-1 antibodies.

Site-Directed Serology for Detection of HIV-1 Infections Using Synthetic Peptides Encompassing the 605–611 Disulfide Loop of gp41

Results described in preceding sections indicated the location of an immunodominant B cell epitope within the 579–611 region of gp41. Other studies identified the immunodominant epitope within a shorter peptide (589–611; fig. 8 [78–84] or within overlapping peptides (596–618 [70] or 596–620 [85]). The propensity to bind anti-gp41 antibodies is increased by intramolecular disulfide bond formation between cysteines at positions 605

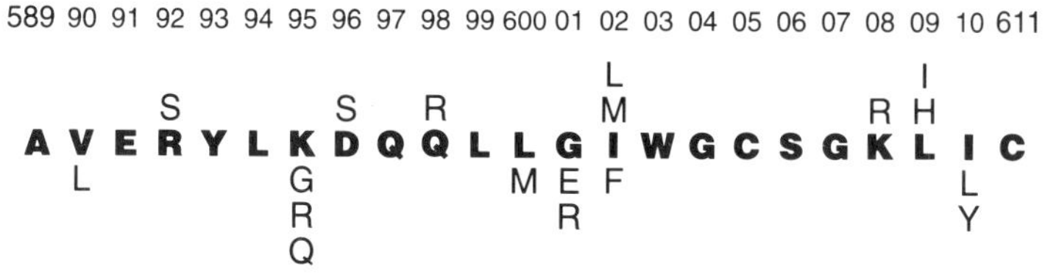

589 90 91 92 93 94 95 96 97 98 99 600 01 02 03 04 05 06 07 08 09 10 611

L I
S S R M R H
A V E R Y L K D Q Q L L G I W G C S G K L I C
L G M E F L
R R Y
Q

Fig. 8. Amino acid sequence encompassing the immunodominant epitope(s) on gp41, and amino acid replacements within this sequence. The consensus sequence corresponding to North American/European HIV-1 isolates is shown in bold. Amino acid replacements frequent in African HIV-1 isolates are shown on top of the sequence. Other replacements are shown below the sequence. The sequences are from references Myers et al. [11] and Horal et al. [70]. Because of inclusion of more than 30 sequences, the variability of the sequence is much greater than that shown in figure 2.

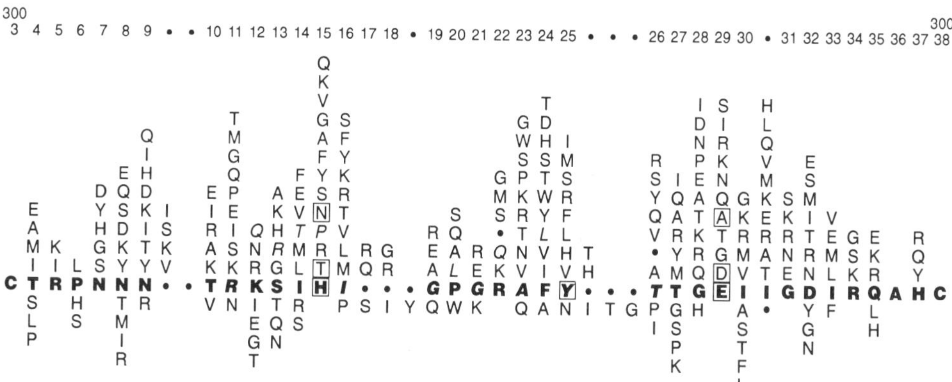

Fig. 9. Amino acid sequence of the V3 loop corresponding to the consensus sequence for North American HIV-1 isolates, and amino acid replacements reported for distinct HIV-1 isolates. Dots correspond to deletions. For consistency, the amino acid residues were numbered according to Ratner et al. [18], published for HIV-1-IIIB. Amino acid replacements shown on top of the consensus sequence are from LaRosa et al. [59] and Wolinsky et al. [71]. Amino acid replacements shown below the consensus sequence are from other reports [72–77]. Amino acid residues characteristic for the Ugandan consensus sequence [73] are shown in italics. Boxed amino acids were consistently found in macrophage-tropic isolates [45, 46]. Because of inclusion of more than 30 sequences, the variability of the sequence is much greater than that shown in figure 2.

and 611, stabilizing a conformation probably near to that of the native glycoprotein [86]. However, other studies indicated that reduction of the disulfide bond did not reduce the capacity of the peptides to serve as antigen for detection of anti-HIV-1 antibodies [83]. Only precise determination of antibody dilution endpoints with reduced and unreduced forms of peptides using a multitude of anti-HIV-1 sera could resolve this apparent discrepancy. Studies with amino acid replacement or deletion set peptides revealed the dominant role of the (604)G–(609)L segment (fig. 8) for antibody binding. A minor fraction of anti-HIV-1 antibodies also reacted with a second site encompassing residues (595)K–(596)D [81, 83, 84].

The immunodominant 589–611 segment of gp41 is not completely conserved among distinct HIV-1 isolates (47.8% variability; fig. 8). The impact of individual amino acid replacements shown in figure 8 on recognition of peptides from the 589–611 region of gp41 has not been systematically investigated. However, selected peptides overlapping this region of gp41 corresponding to North American/European HIV-1 isolates do not cross-react serologically with analogous peptides corresponding to some African HIV-1 isolates [70]. These findings suggest the need to use several peptides from the 589–611 region of distinct HIV-1 isolates as site-specific reagents for detection of anti-HIV-1 antibodies in order to improve the test sensitivity. The choice of peptides to be included in a reagent mixture will depend on the immunological cross-reactivity of the distinct peptides (fig. 8) which can be established only experimentally. Increased sensitivity of site-directed serological assays may be important for the development of simple anti-HIV-1 IgA assays for early diagnosis of perinatal HIV-1 infection [87–89].

Synthetic Peptides as Site-Specific Reagents for Subtyping of Anti-HIV-1 Antibodies

For epidemiological studies, serodiagnosis of HIV-1 infections and the development of anti-HIV-1 vaccines, it is important to understand the amino acid sequence variability of env glycoproteins of viable HIV-1 isolates. This information can be obtained by nucleotide sequencing which is applicable only to a limited number of specimens. Therefore, it seems advantageous to combine sequencing with serological assays permitting distinguishing distinct HIV-1 isolates (i.e. the corresponding antibodies) from each other. Synthetic peptides corresponding to immunodominant continuous B cell epitopes would seem to be best suited for this purpose. Potential candidates for subgrouping of anti-HIV-1 isolates are peptides corresponding to the immunodominant B cell epitope on gp41 with amino acid replacements at positions 601, 602, 608, 609 and 610 (fig. 8). The immunological cross-reactivity between these peptides would have to be established first, followed by selection of the least cross-reactive peptides for anti-HIV-1 subgrouping. Already published data [70] indicate that the L → H replacement at residue 609 leads to considerable changes in antigenic specificity and that anti-HIV-1-positive sera of East African origin may not recognize the consensus sequence peptide corresponding to North American/European HIV-1 isolates.

The homology between sequences corresponding to V3 loops of distinct HIV-1 isolates is only ~ 7% (fig. 9). This offers the opportunity to use selected V3 loop sequences for subtyping of anti-HIV-1. Since anti-V3-specific antibodies predominantly recognize the central portion of the V3 loop (residues ~ 311–326) [19, 59], selected minimally cross-reactive peptides from this region could potentially serve as subtyping reagents. The extent of immunological cross-reactivity or the lack of it can be roughly predicted from amino acid divergence scores distinguishing the peptides from each other [19]. Peptides corresponding to consensus sequences determined for HIV-1 isolates from particular geographical areas might also serve as potential subtyping reagents. Synthetic peptides from the V3 loop have already been used to some extent for subtyping of anti-HIV-1 antibodies [64–67, 76, 77, 90].

Synthetic Peptides as Potential Components of Anti-HIV-1 Vaccines

Only peptides having high antigenic similarity to segments of gp120/gp41 (fig. 6, 7) should be considered as potential components of anti-HIV-1 vaccines. Among these peptides, only those mimicking the V3 loop consistently elicited in immunized animals antibodies efficiently neutralizing sequence-matched HIV-1 [19, 61–63, 91–95]. Other peptides from gp120/gp41 were also reported to elicit virus-neutralizing antibodies [61, 69, 96–99], but their potential application as vaccine components remains to be established. The finding that peptides mimicking the immunodominant V3 loop of gp120 elicit virus-neutralizing antibodies was further extended by the demonstration that: (1) peptides from the V3 loop block the virus-neutralizing activity of polyclonal anti-HIV-1 sera; (2) most of the virus-neutralizing activity of such sera can be adsorbed onto immobilized V3 loop peptides and subsequently eluted in immunopurified form [100, 101], and (3) gp120 deletion mutants lacking V3 loop epitopes were reported not to elicit virus-neutralizing antibodies [102]. Thus, the V3 loop encompasses a principal virus-neutralizing determinant (PND).

MoAbs specific for the PND were also shown to neutralize HIV-1 infectivity [103–106]. Antibodies against the PND were protective in chimpanzees challenged with sequence-matched HIV-1 [69, 107, 108]. The virus-neutralizing and protective activity of PND-specific antibodies can probably be ascribed to their inhibitory activity on HIV-1 entry into cells [42, 43, 109, 110], possibly involving proteolytic cleavage of the V3 loop [111–113].

Notwithstanding the fact that synthetic peptides encompassing the PND sequence may elicit virus-neutralizing and protective antibodies, there are several problems which have to be solved in order to utilize V3 loop peptides as potential vaccine immunogens. These include: (1) incomplete mimicry of the virion V3 loop by synthetic peptides [62]; (2) some cryptotopic features of B cell epitopes on the V3 loop, and/or conformational changes within the V3 loop, resulting from virus assembly [4, 114]; (3) the high variability of the V3 loop (fig. 9), and (4) the generation of virus variants having amino acid replacements (a) within the V3 sequence and (b) outside this sequence but resulting in conformational changes within V3 loops [74–77, 115, 116]. Such replacements may also lead to altered cell tropism of HIV-1 [40–42, 44–46].

Several of the amino acid replacements (fig. 9) result also in the disappearance or appearance of potential N-glycosylation sites (N-X-S/T) at distinct positions of the V3 loop sequence, possibly leading to changes of antigenicity [47].

The most prevalent sequences in the central (319–321) portion of the PND are GPG and GLG (fig. 9). Synthetic peptides from the PND having the G, P, G residues individually deleted had a drastically decreased reactivity with sera from HIV-infected individuals of presumably North American/European origin and with anti-PND-specific MoAbs. Deletions of amino acids preceding or following GPG had a lesser effect. Thus, the GPG sequence is critical for the antigenicity of PND and for virus-neutralizing activity of anti-PND antibodies [104]. Variant viruses with (319)G→D,T,E,W,V,A, (321)G→I,F,R,K,W,L and (320)P→S,Q,V substitutions, generated by in vitro mutagenesis, were reported to be noninfectious [117, 118]. Thus, at least some of the variant HIV-1 sequences may correspond to nonviable or defective viruses, suggesting that the impediment of sequence variability on vaccine development is less serious than implied by figure 9. Therefore, there is hope that immunogens able to elicit antibodies to conserved motifs in PND of distinct HIV-1 isolates, and a combination of a limited number of immunogens cross-reactive with the most prevalent PND sequences, will elicit antibodies neutralizing a majority of HIV-1 isolates and provide protection against infection with these isolates [104, 119–121].

Discontinuous Virus-Neutralizing Epitopes

The majority of virus-neutralizing antibodies (directed against the PND) can be removed by affinity chromatography on peptides derived from the V3 region. The unabsorbed antibodies are also virus-neutralizing but have broader specificity than do anti-PND antibodies, and thus neutralize several HIV-1 isolates [122]. These antibodies do not react and cannot be elicited with denatured gp120, but react with and can be elicited by native gp120 [123, 124], suggesting that they recognize discontinuous epitopes. Such antibodies can also be separated from human anti-HIV-1 by affinity chromatography on columns of CD4-blocked gp120. The group-specific neutralizing antibodies did not absorb to this column, while anti-PND-specific antibodies did [122], indicating that the group-specific neutralizing antibodies were specific for the CD4 attachment site on gp120. These antibodies were reported to recognize discontinuous epitopes within the 349–518 region of gp120 [125]. MoAbs with similar properties were produced by immortalizing B cells from HIV-1-infected individuals [126–128] and by immunization of rodents with gp120 [129, 130]. The discontinuous epitope of one of the MoAbs was disrupted by amino acid replacements at positions 263–264, 375–377, 428 and 477–491 [126]. Attachment of the same MoAbs to gp120 was inhibited by antibodies to the synthetic peptides 458–488 and 438–466, respectively, but not by antibodies to peptides from other regions of gp120 [4]. The attachment of other MoAbs to the CD4-binding site on gp120 was inhibited by antibodies to several peptides overlapping segments 113–142, 254–361 and 425–518 [4], regions which overlap the proposed binding sites for CD4 [131].

A truncated deletion variant of gp120 lacking the 62N- and 20C-terminal residues, as well as residues 127–211 and the V3 loop, and fully retaining the CD4-binding capacity of intact gp120, was recently proposed as an immunogen for eliciting group-specific virus-neutralizing antibodies (fig. 10) [132]. The amino acid sequence of such amputated protein is not conserved within distinct HIV-1 isolates (fig. 2). Therefore, several versions of this protein would have to be included into an immunogen preparation expected to elicit broad protection against infection.

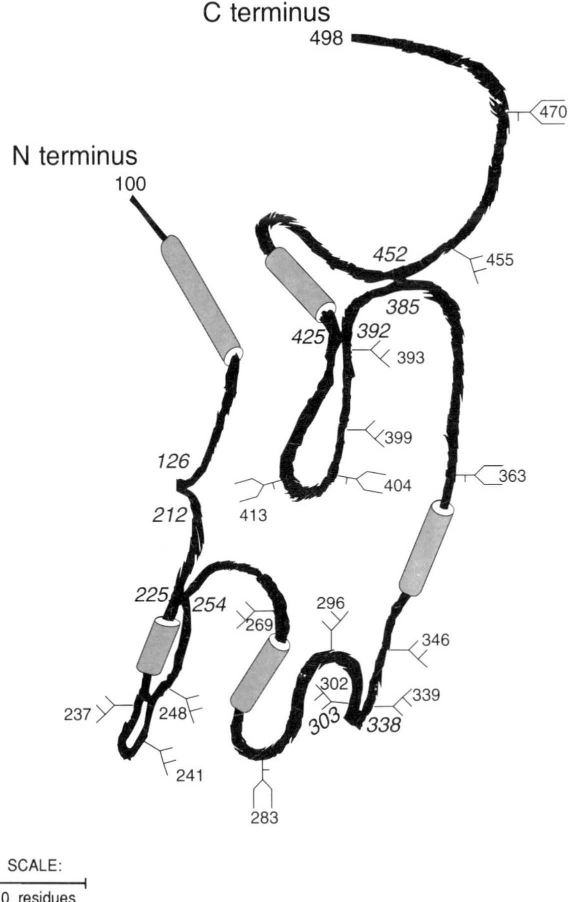

Fig. 10. Schematic representation of unfolded amputated gp120 containing all sites involved in CD4 binding [4, 131, 132] and lacking some of the immunodominant gp120 B cell epitopes (compare with fig. 3, 5 and 6). The V3 loop is replaced by the sequence CRGPC [132].

MoAbs reacting with neotopes generated by oligomerization of gp41 were identified [133], but their reactivity with gp120/gp41 oligomers within the viral envelope was not defined. Antibodies recognizing native gp120/gp41 exposed on cell membranes of infected cells, but not the denatured glycoproteins, appearing early in the course of HIV-1 infection [134], also remain to be characterized.

Molecular Mimicry of gp120/gp41 B Cell Epitopes

Comparisons of the gp120/gp41 sequence with proteins in computer data banks revealed significant sequence homologies between the immunoglobulin superfamily and the 49–92, 261–270, 275–287, 425–445, 590–599 and 831–837 segments of gp120/gp41 [29, 30, 32, 135–137]. In accordance with these results, apparent immunological cross-reactivities between gp120/gp41 and HLA-DR α- and β-chains were observed experimentally. All these partial sequence homology regions correspond to conserved portions of gp120/gp41 or to fragments with mostly conservative amino acid replacements. The molecular mimicry between HIV-1 and HLA class II antigens may potentially lead to the elicitation of autoantibodies in infected individuals leading to impaired immune responses against newly encountered antigens. The sequence homology regions in gp120/gp41 correspond to segments having a low rank in the hierarchy of B cell epitopes (fig. 3–6). This seems to agree with the general conclusion that epitopes involved in autoimmunity do not correspond to dominant epitopes [138]. Discontinuous epitopes on gp120/gp41 cross-reactive with epitopes on members of the immunoglobulin superfamily are likely to play a more important role in eliciting autoimmunity following HIV-1 infection than do continuous epitopes.

Partial sequence homology and immunological cross-reactivity between interleukin-2 and the C terminus of gp41 were also found [139]. Partial sequence homology between neuroleukin and the 245–289 region of gp120 was described [31], and MoAbs to the immunodominant gp41 epitope were reported to bind to a component of astrocytes [28]. These observations may possibly explain neuronal dysfunctions observed in some patients with AIDS.

Human endogenous HIV-1-related sequences were detected by nucleic acid hybridization [140]. Their deduced amino acid sequences share regions of homology with the 507–520 and 736–754 segments of gp120/gp41.

Conclusions

B cell antigenic site mapping of HIV-1 glycoproteins gp120/gp41 using synthetic peptides and antipeptide antisera has contributed to the design of diagnostic reagents for detecting anti-HIV-1 antibodies and to the development of strategies for passive and active immunization against HIV-1 infections. Several problems related to the characterization of B cell epitopes on HIV-1 env glycoproteins remain to be solved. These include: (1) immunological analysis of the impact of amino acid replacements distinguishing HIV-1 subtypes from each other on the antigenicity and immunogenicity of gp120/gp41 and of segments derived from them; (2) characterization of epitopes on the envelope of intact virions, including the definition of cryptotopes and neotopes and the elucidation of the role of carbohydrate chains in limiting the accessibility of B cell epitopes on intact virions, and (3) characterization of discontinuous epitopes contributing to both immunity to HIV-1 and to autoimmune responses, which may play a role in the pathogenesis of HIV-1 infections. These problems may probably be addressed by analysis of viral mutants with MoAbs, by competitive binding assays with site-specific mini-antibodies (= dimeric F_v fragments) [141] and by X-ray crystallography (methods 1, 7, 8 and 10; table 1).

References

1 Putney SD, Montelaro RC: Lentiviruses; in Van Regenmortel MHV, Neurath AR (eds): Immunochemistry of Viruses. II. The Basis for Serodiagnosis and Vaccines. Amsterdam, Elsevier, 1990, pp 307–344.
2 Gelderblom HR: Assembly and morphology of HIV: Potential effect of structure on viral function. AIDS 1991;5:617–638.
3 Gallaher WR, Ball JM, Garry RF, Griffin MC, Montelaro RC: A general model for the transmembrane proteins of HIV and other retroviruses. AIDS Res Hum Retroviruses 1989;5:431–440.
4 Neurath AR, Strick N, Jiang S: Synthetic peptides and anti-peptide antibodies as probes to study inter-domain interactions involved in virus assembly: The envelope of the human immunodeficiency virus (HIV-1). Virology, in press.
5 Dedera D, Gu R, Ratner L: Conserved cysteine residues in the human immunodeficiency virus type 1 transmembrane envelope protein are essential for precursor envelope cleavage. J Virol 1992;66:1207–1209.
6 Syu W-J, Lee W-R, Du B, Yu Q-C, Essex M, Lee T-H: Role of conserved gp41 cysteine residues in the processing of human immunodeficiency virus envelope precursor and viral infectivity. J Virol 1991;65:6349–6352.
7 Travis BM, Dykers TI, Hewgill D, Ledbetter J, Tsu TT, Hu S-L, Lewis JB: Functional roles of the V3 hypervariable region of HIV-1 gp160 in the processing of gp160 and in the formation of syncytia in CD4+ cells. Virology 1992;186:313–317.
8 Wang J, Yan Y, Garrett TPJ, Liu J, Rodgers DW, Garlick RL, Tarr GE, Husain Y, Reinherz EL, Harrison SC: Atomic structure of a fragment of human CD4 containing two immunoglobulin-like domains. Nature 1990;348:411–418.
9 Moore JP, McKeating JA, Weiss RA, Sattentau QJ: Dissociation of gp120 from HIV-1 virions induced by soluble CD4. Science 1990;250:1139–1142.
10 Wu TT, Kabat EA: An analysis of the sequences of the variable regions of Bence-

Jones proteins and myeloma light chains and their implications for antibody complementarity. J Exp Med 1970;132:211–250.

11 Myers G, Berzofsky JA, Korber B, Smith RF (eds): Human Retroviruses and AIDS 1991. Los Alamos, Los Alamos National Laboratory, 1991.

12 Turner S, Tizard R, DeMarinis J, Pepinsky RB, Zullo J, Schooley R, Fisher R: Resistance of primary isolates of human immunodeficiency virus type 1 to neutralization by soluble CD4 is not due to lower affinity with the viral envelope glycoprotein gp120. Proc Natl Acad Sci USA 1992;89:1335–1339.

13 Moore JP, McKeating JA, Huang Y, Ashkenazi A, Ho DD: Virions of primary human immunodeficiency virus type 1 isolates resistant to soluble CD4 (sCD4) neutralization differ in sCD4 binding and glycoprotein gp120 retention from sCD4-sensitive isolates. J Virol 1992;66:235–243.

14 Brighty DW, Rosenberg M, Chen ISY, Ivey-Hoyle M: Envelope proteins from clinical isolates of human immunodeficiency virus type 1 that are refractory to neutralization by soluble CD4 possess high affinity for the CD4 receptor. Proc Natl Acad Sci USA 1991;88:7802–7805.

15 Ashkenazi A, Smith DH, Marsters SA, Riddle L, Gregory TJ, Ho DD, Capon DJ: Resistance of primary isolates of human immunodeficiency virus type 1 to soluble CD4 is independent of CD4-rgp120 binding affinity. Proc Natl Acad Sci USA 1991; 88:7056–7060.

16 Neurath AR, Strick N, Taylor P, Rubinstein P, Stevens CE: Search for epitope-specific antibody responses to the human immunodeficiency virus (HIV-1) envelope glycoproteins signifying resistance to disease development. AIDS Res Hum Retroviruses 1990;6:1183–1191.

17 Leonard CK, Spellman MW, Riddle L, Harris RJ, Thomas JN, Gregory TJ: Assignment of intrachain disulfide bonds and characterization of potential glycosylation sites of the type 1 recombinant human immunodeficiency virus envelope glycoprotein (gp120) expressed in Chinese hamster ovary cells. J Biol Chem 1990;265: 10373–10382.

18 Ratner L, Haseltine W, Patarca R, Livak KJ, Starcich B, Josephs SF, Doran ER, Rafalski JA, Whitehorn EA, Baumeister K, Ivanoff L, Petteway SR Jr, Pearson ML, Lautenberger JA, Papas TS, Ghrayeb J, Chang NT, Gallo RC, Wong-Staal F: Complete nucleotide sequence of the AIDS virus, HTLV-III. Nature 1985;313:277–284.

19 Neurath AR, Strick N: Confronting the hypervariability of an immunodominant epitope eliciting virus neutralizing antibodies from the envelope glycoprotein of the human immunodeficiency virus type 1 (HIV-1). Mol Immunol 1990;27:539–549.

20 Lee W-H, Yu X-F, Syu W-J, Essex M, Lee T-H: Mutational analysis of conserved N-linked glycosylation sites of human immunodeficiency virus type 1 gp41. J Virol 1992;66:1799–1803.

21 Dedera DA, Gu R, Ratner L: Role of asparagine-linked glycosylation in human immunodeficiency virus type 1 transmembrane envelope function. Virology 1992; 187:377–382.

22 Dirckx L, Lindemann D, Ette R, Manzoni C, Moritz D, Mous J: Mutation of conserved *N*-glycosylation sites around the CD4-binding site of human immunodeficiency virus type 1 GP120 affects viral infectivity. Virus Res 1990;18:9–20.

23 Hansen J-ES, Nielsen C, Arendrup M, Olofsson S, Mathiesen L, Nielsen JO, Clausen

H: Broadly neutralizing antibodies targeted to mucin-type carbohydrate epitopes of human immunodeficiency virus. J Virol 1991;65:6461–6467.

24 Harouse JM, Bhat S, Spitalnik SL, Laughlin M, Stefano K, Silberberg DH, Gonzalez-Scarano F: Inhibition of entry of HIV-1 in neural cell lines by antibodies against galactosyl ceramide. Science 1991;253:320–323.

25 Neurath AR, Haberfield P, Joshi B, Hewlett IK, Strick N, Jiang S: Rapid prescreening for antiviral agents against HIV-1 based on their inhibitory activity in site-directed immunoassays. I. The V3 loop of gp120 as target. Antiviral Chem Chemother 1991;2:303–312.

26 Neurath AR, Strick N, Haberfield P, Jiang S: Rapid prescreening for antiviral agents against HIV-1 based on their inhibitory activity in site-directed immunoassays. II. Porphyrins reacting with the V3 loop of gp120. Antiviral Chem Chemother, in press.

27 Jehuda-Cohen T, Slade BA, Powell JD, Villinger F, De B, Folks TM, McClure HM, Sell KW, Ahmed-Ansari A: Polyclonal B-cell activation reveals antibodies against human immunodeficiency virus type 1 (HIV-1) in HIV-1-seronegative individuals. Proc Natl Acad Sci USA 1990;87:3972–3976.

28 Yamada M, Zurbriggen A, Oldstone MBA, Fujinami RS: Common immunologic determinant between human immunodeficiency virus type 1 gp41 and astrocytes. J Virol 1991;65:1370–1376.

29 Golding H, Robey FA, Gates FT III, Linder W, Beining PR, Hoffman T, Golding B: Identification of homologous regions in human immunodeficiency virus I gp41 and human MHC class II beta 1 domain. J Exp Med 1988;167:914–923.

30 Golding H, Shearer GM, Hillman K, Lucas P, Manischewitz J, Zajac RA, Clerici M, Gress RE, Boswell RN, Golding B: Common epitope in human immunodeficiency virus (HIV) I-gp41 and HLA class II elicits immunosuppressive autoantibodies capable of contributing to immune dysfunction in HIV-1-infected individuals. J Clin Invest 1989;83:1430–1435.

31 Lee MR, Ho DD, Gurney ME: Functional interaction and partial homology between human immunodeficiency virus and neuroleukin. Science 1987;237:1047–1051.

32 Maddon PJ, Dalgleish AG, McDougal JS, Clapham PR, Weiss RA, Axel R: The T4 gene encodes the AIDS virus receptor and is expressed in the immune system and the brain. Cell 1986;47:333–338.

33 Reiher WE III, Blalock JE, Brunck TK: Sequence homology between acquired immunodeficiency syndrome virus envelope protein and interleukin 2. Proc Natl Acad Sci USA 1986;83:9188–9192.

34 Weigent DA, Hoeprich PD, Bost KL, Brunck TK, Reiher WE III, Blalock JE: The HTLV-III envelope protein contains a hexapeptide homologous to a region of interleukin-2 that binds to the interleukin-2 receptor. Biochem Biophys Res Commun 1986;139:367–374.

35 Veljković V, Metlaš R: Sequence similarity between human immunodeficiency virus type 1 envelope protein (gp120) and human proteins. A new hypothesis on protective antibody production. Immunol Lett 1990;26:193–196.

36 Ameisen JC, Capron A: Cell dysfunction and depletion in AIDS: The programmed cell death hypothesis. Immunol Today 1991;12:102–105.

37 Groux H, Torpier G, Monté D, Mouton Y, Capron A, Ameisen JC: Activation-induced death by apoptosis in CD4+ T cells from human immunodeficiency virus-infected asymptomatic individuals. J Exp Med 1992;175:331–340.

38 Imberti L, Sottini A, Bettinardi A, Puoti M, Primi D: Selective depletion in HIV infection of T cells that bear specific T cell receptor V_β sequences. Science 1991;254: 860–862.

39 Jiang S, Lin K, Neurath AR: Enhancement of human immunodeficiency virus type 1 infection by antisera to peptides from the envelope glycoprotein gp120/gp41. J Exp Med 1991;174:1557–1563.

40 Takeuchi Y, Akutsu M, Murayama K, Shimizu N, Hoshino H: Host range mutant of human immunodeficiency virus type 1: Modification of cell tropism by a single point mutation at the neutralization epitope in the *env* gene. J Virol 1991;65:1710–1718.

41 Cann AJ, Churcher MJ, Boyd M, O'Brien W, Zhao J-Q, Zack J, Chen ISY: The region of the envelope gene of human immunodeficiency virus type 1 responsible for determination of cell tropism. J Virol 1992;66:305–309.

42 Cheng-Mayer C, Shioda T, Levy JA: Host range, replicative, and cytopathic properties of human immunodeficiency virus type 1 are determined by very few amino acid changes in *tat* and gp120. J Virol 1991;65:6931–6941.

43 de Jong J-J, Goudsmit J, Keulen W, Klaver B, Krone W, Tersmette M, de Ronde A: Human immunodeficiency virus type 1 clones chimeric for the envelope V3 domain differ in syncytium formation and replication capacity. J Virol 1992;66:757–765.

44 Hwang SS, Boyle TJ, Lyerly HK, Cullen BR: Identification of the envelope V3 loop as the primary determinant of cell tropism in HIV-1. Science 1991;253:71–74.

45 Westervelt P, Trowbridge DB, Epstein LG, Blumberg BM, Li Y, Hahn BH, Shaw GM, Price RW, Ratner L: Macrophage tropism determinants of human immunodeficiency virus type 1 in vivo. J Virol 1992;66:2577–2582.

46 Sharpless NE, O'Brien WA, Verdin E, Kufta CV, Chen ISY, Dubois-Dalcq M: Human immunodeficiency virus type 1 tropism for brain microglial cells is determined by a region of the *env* glycoprotein that also controls macrophage tropism. J Virol 1992;66:2588–2593.

47 Klenk HD: Influence of glycosylation on antigenicity of viral proteins; in Van Regenmortel MHV, Neurath AR (eds): Immunochemistry of Viruses. II. The Basis for Serodiagnosis and Vaccines. Amsterdam, Elsevier, 1990, pp 25–37.

48 Van Regenmortel MHV: The structure of viral epitopes; in Van Regenmortel MHV, Neurath AR (eds): Immunochemistry of Viruses. II. The Basis for Serodiagnosis and Vaccines. Amsterdam, Elsevier, 1990, pp 1–24.

49 Barlow DJ, Edwards MS, Thornton JM: Continuous and discontinuous protein antigenic determinants. Nature 1986;322:747–748.

50 Van Regenmortel MHV: Molecular dissection of protein antigens; in Van Regenmortel MHV (ed): Structure of Antigens. Boca Raton, CRC Press, 1992, vol 1, pp 1–27.

51 Jemmerson R, Paterson Y: Mapping epitopes on a protein antigen by the proteolysis of antigen-antibody complexes. Science 1986;232:1001–1004.

52 de la Cruz VF, Lal AA, McCutchan TF: Immunogenicity and epitope mapping of foreign sequences via genetically engineered filamentous phage. J Biol Chem 1988; 263:4318–4322.

53 Nunberg JH, Rodgers G, Gilbert JH, Snead RM: Method to map antigenic determinants recognized by monoclonal antibodies. Localization of a determinant of virus neutralization on the feline leukemia virus envelope protein gp70. Proc Natl Acad Sci USA 1984;81:3675–3679.

54 Tsunetsugu-Yokota Y, Tatsumi M, Robert V, Devaux C, Spire B, Chermann J-C,

Hirsch I: Expression of an immunogenic region of HIV by a filamentous bacteriophage vector. Gene 1991;99:261–265.

55 Goodenow M, Huet T, Surin W, Kwok S, Sninksy J, Wain-Hobson S: HIV-1 isolates are rapidly evolving quasispecies: Evidence for viral mixtures and preferred nucleotide substitutions. J AIDS 1989;2:344–352.

56 Meyerhans A, Cheynier R, Albert J, Seth M, Kwok S, Snisky J, Morfeldt-Manson L, Asjo B, Wain-Hobson S: Temporal fluctuations in HIV quasispecies in vivo are not reflected by sequential HIV isolations. Cell 1989;58:901–910.

57 Nara PL, Smit L, Dunlop N, Hatch W, Merges M, Waters D, Kelliher J, Gallo RC, Fischinger PJ, Goudsmit J : Emergence of viruses resistant to neutralization by V3-specific antibodies in experimental human immunodeficiency virus type 1 IIIB infection of chimpanzees. J Virol 1990;64:3779–3791.

58 Wain-Hobson S, Sonigo P, Danos O, Cole S, Alizon M: Nucleotide sequence of the AIDS virus, LAV. Cell 1985;40:9–17.

59 LaRosa GJ, Davide JP, Weinhold K, Waterbury JA, Profy AT, Lewis JA, Langlois AJ, Dreesman GR, Boswell RN, Shadduck P, Holley LH, Karplus M, Bolognesi DP, Matthews TJ, Emini EA, Putney SD: Conserved sequence and structural elements in the HIV-1 principal neutralizing determinant. Science 1990;249:932–935.

60 Sanchez-Pescador R, Power MD, Barr PJ, Steiner KS, Stempien MM, Brown-Shimer SL, Gee WW, Renard A, Randolph A, Levy JA, Dina D, Luciw PA: Nucleotide sequence and expression of an AIDS-associated retrovirus (ARV-2). Science 1985; 227:484–491.

61 Neurath AR, Strick N, Lee ESY: B cell epitope mapping of human immunodeficiency virus envelope glycoproteins with long (19- to 36-residue) synthetic peptides. J Gen Virol 1990;71:85–95.

62 Neurath AR, Jiang S, Strick N, Kolbe H, Kieny M-P, Muchmore E, Girard M: Antibody responses of chimpanzees immunized with synthetic peptides corresponding to full-length V3 hypervariable loops of HIV-1 envelope glycoproteins. AIDS Res Hum Retroviruses 1991;7:813–823.

63 Neurath AR, Strick N, Fields R, Jiang S: Peptides mimicking selected disulfide loops in HIV-1 gp120, other than V3, do not elicit virus-neutralizing antibodies. AIDS Res Hum Retroviruses 1991;7:657–661.

64 Devash Y, Matthews TJ, Drummond JE, Javaherian K, Waters DJ, Arthur LO, Blattner WA, Rusche JR: C-terminal fragments of gp120 and synthetic peptides from five HTLV-III-MN isolates in infected individuals. AIDS Res Hum Retroviruses 1990;6:307–316.

65 Goudsmit J, Zwart G, Bakker M, Smit L, Back N, Epstein L, Kuiken CX, D'Amaro J, De Wolf F: Antibody recognition of amino acid divergence within an HIV-1 neutralization epitope. Res Virol 1989;140:419–436.

66 Goudsmit J, Zwart G, de Jong J, Wolfs T, de Ronde A, Nara PL: Antigenic variation in neutralization domains of HIV-1 strains circulating during the current AIDS epidemic: Key to rational vaccine design; in Brown F, Chanock RM, Ginsburg H, Lerner RA (eds): Vaccines 90: Modern Approaches to New Vaccines Including Prevention of AIDS. Cold Spring Harbor, Cold Spring Harbor Laboratory Press, 1990, pp 291–295.

67 Zwart G, de Jong JJ, Wolfs T, van der Hoek L, Smit L, de Ronde A, Tersmette M, Nara P, Goudsmit J: Predominance of HIV-1 serotype distinct from LAV-1/HTLV-IIIB. Lancet 1990;335:474.

68 Getzoff ED, Tainer JA, Lerner RA: The chemistry and mechanism of antibody binding to protein antigens; in Dixon FJ (ed): Advances in Immunology. San Diego, Academic Press, 1988, vol 43, pp 1–98.
69 Berman PW, Gregory TJ, Riddle L, Nakamura GR, Champe MA, Porter JP, Wurm FM, Hershberg RD, Cobb EK, Eichberg JW: Protection of chimpanzees from infection by HIV-1 after vaccination with recombinant glycoprotein gp120 but not gp160. Nature 1990;345:622–625.
70 Horal P, Svennerholm B, Jeansson S, Rymo L, Hall WW, Vahlne A: Continuous epitopes of the human immunodeficiency virus type 1 (HIV-1) transmembrane glycoprotein and reactivity of human sera to synthetic peptides representing various HIV-1 isolates. J Virol 1991;65:2718–2723.
71 Wolinsky SM, Wike CM, Korber BTM, Hutto C, Parks WP, Rosenblum LL, Kunstman KJ, Furtado MR, Muñoz JL: Selective transmission of human immunodeficiency virus type-1 variants from mothers to infants. Science 1992;255:1134–1137.
72 Hattori T, Shiozaki K, Eda Y, Tokiyoshi S, Matsushita S, Inaba H, Fujimaki M, Meguro T, Yamada K, Honda M, Nishikawa K, Takatsuki K: Characteristics of the principal neutralizing determinant of HIV-1 prevalent in Japan. AIDS Res Hum Retroviruses 1991;7:825–830.
73 Oram JD, Downing RG, Roff M, Serwankambo N, Clegg JCS, Featherstone ASR, Booth JC: Sequence analysis of the V3 loop regions of the *env* genes of Ugandan human immunodeficiency proviruses. AIDS Res Hum Retroviruses 1991;7:605–614.
74 Wahlberg J, Albert J, Lundeberg J, von Gegerfelt A, Broliden K, Utter G, Fenyö EM, Uhlén M: Analysis of the V3 loop in neutralization-resistant human immunodeficiency virus type 1 variants by direct solid-phase DNA sequencing. AIDS Res Hum Retroviruses 1991;7:983–990.
75 Wolfs TFW, de Jong J-J, van den Berg H, Tijnagel JMGH, Krone WJA, Goudsmit J: Evolution of sequences encoding the principal neutralization epitope of human immunodeficiency virus 1 is host dependent, rapid, and continuous. Proc Natl Acad Sci USA 1990;87:9938–9942.
76 Wolfs TFW, Zwart G, Bakker M, Valk M, Kuiken CL, Goudsmit J: Naturally occurring mutations within HIV-1 V3 genomic RNA lead to antigenic variation dependent on a single amino acid substitution. Virology 1991;185:195–205.
77 Zwart G, Langedijk H, van der Hoek L, de Jong J-J, Wolfs TFW, Ramautarsing C, Bakker M, de Ronde A, Goudsmit J: Immunodominance and antigenic variation of the principal neutralization domain of HIV-1. Virology 1991;181:481–489.
78 Chiodi F, von Gegerfeldt A, Albert J, Fenyö EM, Gaines H, von Sydow M, Biberfeld G, Parks E, Norrby E: Site-directed ELISA with synthetic peptides representing the HIV transmembrane glycoprotein. J Med Virol 1987;23:1–9.
79 Chiodi F, Mathiesen T, Albert J, Parks E, Norrby E, Wahren B: IgG subclass responses to a transmembrane protein (gp41) peptide in HIV infection. J Immunol 1989;142:3809–3814.
80 Norrby E: Human immunodeficiency virus antibody responses determined by site-directed serology. Intervirology 1990;31:315–326.
81 Norrby E, Biberfeld G, Johnson PR, Parks DE, Houghten RA, Lerner RA: The chemistry of site-directed serology for HIV infections. AIDS Res Hum Retroviruses 1989;5:487–493.

82 Norrby E, Biberfeld G, Chiodi F, von Gegerfeldt A, Nauclér A, Parks E, Lerner R: Discrimination between antibodies to HIV and to related retroviruses using site-directed serology. Nature 1987;329:248–2560.
83 Norrby E, Parks DE, Utter G, Houghten RA, Lerner RA: Immunochemistry of the dominating antigenic region Ala^{582} to Cys^{604} in the transmembranous protein of simian and human immunodeficiency virus. J Immunol 1989;143:3602–3608.
84 Mathiesen T, Chiodi F, Broliden P-A, Albert J, Houghten RA, Utter G, Wahren B, Norrby E: Analysis of a subclass-restricted HIV-1 gp41 epitope by omission peptides. Immunology 1989;67:1–7.
85 Närvänen A, Korkolainen M, Suni J, Korpela J, Kontio S, Partanen P, Vaheri A, Huhtala M-L: Synthetic *env* gp41 peptide as a sensitive and specific diagnostic reagent in different stages of human immunodeficiency virus type 1 infection. J Med Virol 1988;26:111–118.
86 Oldstone MBA, Tishon A, Lewicki H, Dyson HJ, Feher VA, Assa-Munt N, Wright PE: Mapping the anatomy of the immunodominant domain of the human immunodeficiency virus gp41 transmembrane protein: Peptide conformation analysis using monoclonal antibodies and proton nuclear magnetic resonance spectroscopy. J Virol 1991;65:1727–1734.
87 Landesman S, Weiblen B, Mendez H, Willoughby A, Goedert JJ, Rubinstein A, Minkoff H, Moroso G: Clinical utility of HIV-IgA immunoblot assay in the early diagnosis of perinatal HIV infection. JAMA 1991;266:3443–3446.
88 Quinn TC, Kline RL, Halsey N, Hutton N, Ruff A, Butz A, Boulos R, Modlin JF: Early diagnosis of perinatal HIV infection by detection of viral-specific IgA antibodies. JAMA 1991;266:3439–3446.
89 Weiblen BJ, Lee FK, Cooper ER, Landesman SH, McIntosh K, Harris J-AS, Nesheim S, Mendez H, Pelton SI, Nahmias AJ, Hoff R: Early diagnosis of HIV infection in infants by detection of IgA HIV antibodies. Lancet 1990:335:988–990.
90 Carrow EW, Vujcic LK, Glass WL, Seamon KB, Rastogi SC, Hendry RM, Boulos R, Nzila N, Quinnan GV Jr: High prevalence of antibodies to the gp120 V3 region principal neutralizing determinant of $HIV\text{-}1_{MN}$ in sera from Africa and the Americas. AIDS Res Hum Retroviruses 1991;7:831–838.
91 Neurath AR, Strick N, Kolbe H, Kieny M-P, Girard M, Jiang S: Confronting the hypervariability of an immunodominant epitope eliciting virus neutralizing antibodies from the envelope glycoprotein of the human immunodeficiency virus type 1 (HIV-1). II. Synthetic peptides linked to HIV-1 carrier proteins *gag* and *nef*. Mol Immunol 1991;28:965–973.
92 Nardelli B, Lu Y-A, Shiu DR, Delpierre-Defoort C, Profy AT, Tam JP: A chemically defined synthetic vaccine model for HIV-1. J Immunol 1992;148:914–920.
93 Wang CY, Looney DJ, Li ML, Walfield AM, Ye J, Hosein B, Tam JP, Wong-Staal F: Long-term high-titer neutralizing activity induced by octameric synthetic HIV-1 antigen. Science 1991;254:285–288.
94 Hart MK, Palker TJ, Matthews TJ, Langlois AJ, Lerche NW, Martin ME, Scearce RM, McDanal C, Bolognesi DP, Haynes BF: Synthetic peptides containing T and B cell epitopes from human immunodeficiency virus envelope gp120 induce anti-HIV proliferative responses and high titers of neutralizing antibodies in rhesus monkeys. J Immunol 1990;145:2677–2685.
95 Langedijk JPM, Back NKT, Durda PJ, Goudsmit J, Meloen RH: Neutralizing

activity of anti-peptide antibodies against the principal neutralization domain of human immunodeficiency virus type 1. J Gen Virol 1991;72:2519–2526.

96 Vahlne A, Horal P, Eriksson K, Jeansson S, Rymo L, Hedström K-G, Czerkinsky C, Holmgren J, Svennerholm B: Immunizations of monkeys with synthetic peptides disclose conserved areas on gp120 of human immunodeficiency virus type 1 associated with cross-neutralizing antibodies and T-cell recognition. Proc Natl Acad Sci USA 1991;88:10744–10748.

97 Chanh TC, Dreesman GR, Kanda P, Linette GP, Sparrow JT, Ho DD, Kennedy RC: Induction of anti-HIV neutralizing antibodies by synthetic peptides. EMBO J 1986; 5:3065–3071.

98 Dalgleish AG, Chanh TC, Kennedy RC, Kanda P, Clapham PR, Weiss RA: Neutralization of diverse HIV-1 strains by monoclonal antibodies raised against a gp41 synthetic peptide. Virology 1988;165:209–215.

99 Ho DD, Kaplan JC, Rackauskas IE, Gurney ME: Second conserved domain of gp120 is important for HIV infectivity and antibody neutralization. Science 1988;239: 1021–1023.

100 Profy AT, Salinas PA, Eckler LI, Dunlop NM, Nara PL, Putney SD: Epitopes recognized by the neutralizing antibodies of an HIV-1-infected individual. J Immunol 1990;144:4641–4647.

101 Fauci AS, Gallo RC, Koenig S, Salk J, Purcell RH: NIH Conference. Development and evaluation of a vaccine for human immunodeficiency virus (HIV) infection. Ann Intern Med 1989;110:373–385.

102 Javaherian K, Langlois AJ, McDanal C, Ross KL, Eckler LI, Jellis CL, Profy AT, Rusche JR, Bolognesi DP, Putney SD, Matthews TJ: Principal neutralizing domain of the human immunodeficiency virus type 1 envelope protein. Proc Natl Acad Sci USA 1989;86:6768–6772.

103 Laman JD, Schellekens MM, Abacioglu YH, Lewis GK, Tersmette M, Fouchier RAM, Langedijk JPM, Claasen E, Boersma WJA: Variant-specific monoclonal and group-specific polyclonal human immunodeficiency virus type 1 neutralizing antibodies raised with synthetic peptides from the gp120 third variable domain. J Virol 1992;66:1823–1831.

104 Broliden PA, Mäkitalo B, Åkerblom L, Rosen J, Broliden K, Utter G: Identification of amino acids in the V3 region of gp120 critical for virus neutralization by human HIV-1-specific antibodies. Immunology 1991;73:371–376.

105 Gorny MK, Xu J-Y, Gianakakos V, Karwowska S, Williams C, Sheppard HW, Hanson CV, Zolla-Pazner S: Production of site-selected neutralizing human monoclonal antibodies against the third variable domain of the human immunodeficiency virus type 1 envelope glycoprotein. Proc Natl Acad Sci USA 1991;88:3238–3242.

106 Scott CF Jr, Silver S, Profy AT, Putney SD, Langlois A, Weinhold K, Robinson JE: Human monoclonal antibody that recognizes the V3 region of human immunodeficiency virus gp120 and neutralizes the human T-lymphotropic virus type III_{MN} strain. Proc Natl Acad Sci USA 1990;87:8597–8601.

107 Emini EA, Schleif WA, Nunberg JH, Conley AJ, Eda Y, Tokiyoshi S, Putney SD, Matsushita S, Cobb KE, Jett CM, Eichberg JW, Murthy KK: Prevention of HIV-1 infection in chimpanzees by gp120 V3 domain-specific monoclonal antibody. Nature 1992;355:728–730.

108 Girard M, Kieny M-P, Pinter A, Barre-Sinoussi F, Nara P, Kolbe H, Kusumi K, Chaput A, Reinhart T, Muchmore E, Ronco J, Kaczorek M, Gomard E, Gluckman

JC, Fultz PN: Immunization of chimpanzees confers protection against challenge with human immunodeficiency virus. Proc Natl Acad Sci USA 1991;88:542–546.

109 Page KA, Stearns SM, Littman DR: Analysis of mutations in the V3 domain of gp160 that affect fusion and infectivity. J Virol 1992;66:524–533.

110 Freed EO, Delwart EL, Buchschacher GL Jr, Panganiban AT: A mutation in the human immunodeficiency virus type 1 transmembrane glycoprotein gp41 dominantly interferes with fusion and infectivity. Proc Natl Acad Sci USA 1992;89:70–74.

111 Clements GJ, Price-Jones MJ, Stephens PE, Sutton C, Schulz TF, Clapham PR, McKeating JA, McClure MO, Thomson S, Marsh M, Kay J, Weiss RA, Moore JP: The V3 loops of the HIV-1 and HIV-2 surface glycoproteins contain proteolytic cleavage sites: A possible function in viral fusion? AIDS Res Hum Retroviruses 1991; 7:3–16.

112 Kido H, Fukutomi A, Katunuma N: Tryptase TL_2 in the membrane of human T4+ lymphocytes is a novel binding protein of the V3 domain of HIV-1 envelope glycoprotein gp120. FEBS Lett 1991;286:233–236.

113 Murakami T, Hattori T, Takatsuki K: A principal neutralizing domain of human immunodeficiency virus type 1 interacts with proteinase-like molecule(s) at the surface of Molt-4 clone 8 cells. Biochim Biophys Acta 1991;1079:279–284.

114 Sattentau QJ, Moore JP: Conformational changes induced in the human immunodeficiency virus envelope glycoprotein by soluble CD4 binding. J Exp Med 1991;174: 407–415.

115 Goudsmit J, Kuiken CL, Nara PL: Linear versus conformational variation of V3 neutralization domains of HIV-1 during experimental and natural infection. AIDS 1989;3:S119–S123.

116 Ivanoff LA, Looney DJ, McDanal C, Morris JF, Wong-Staal F, Langlois AJ, Petteway SR Jr, Matthews TJ: Alteration of HIV-1 infectivity and neutralization by a single amino acid replacement in the V3 loop domain. AIDS Res Hum Retroviruses 1991;7:595–603.

117 Ivanoff LA, Dubay JW, Morris JF, Roberts SJ, Gutshall L, Sternberg EJ, Hunter E, Matthews TJ, Petteway SR Jr: V3 loop region of the HIV-1 gp120 envelope protein is essential for virus infectivity. Virology 1992;187:423–432.

118 Grimaila RJ, Fuller BA, Rennert PD, Nelson MB, Hammarskjöld M-L, Potts B, Murray M, Putney SD, Gray G: Mutations in the principal neutralization determinant of human immunodeficiency virus type 1 affect syncytium formation, virus infectivity, growth kinetics, and neutralization. J Virol 1992;66:1875–1883.

119 Javaherian K, Langlois AJ, LaRosa GJ, Profy AT, Bolognesi DP, Herlihy WC, Putney SD, Matthews TJ: Broadly neutralizing antibodies elicited by the hypervariable neutralizing determinant of HIV-1. Science 1990;250:1590–1593.

120 Ohno T, Terada M, Yoneda Y, Shea KW, Chambers RF, Stroka DM, Nakamura M, Kufe DW: A broadly neutralizing monoclonal antibody that recognizes the V3 region of human immunodeficiency virus type 1 glycoprotein gp120. Proc Natl Acad Sci USA 1991;88:10726–10729.

121 Holley LH, Goudsmit J, Karplus M: Prediction of optimal peptide mixtures to induce broadly neutralizing antibodies to human immunodeficiency virus type 1. Proc Natl Acad Sci USA 1991;88:6800–6804.

122 Kang C-Y, Nara P, Chamat S, Caralli V, Ryskamp T, Haigwood N, Newman R, Köhler H: Evidence for non-V3-specific neutralizing antibodies that interfere with

gp120/CD4 binding in human immunodeficiency virus 1-infected humans. Proc Natl Acad Sci USA 1991;88:6171–6175.

123 Haigwood NL, Nara PL, Brooks E, van Nest GA, Ott G, Higgins KW, Dunlop N, Scandella CJ, Eichberg JW, Steimer KS: Native but not denatured recombinant human immunodeficiency virus type 1 gp120 generates broad-spectrum neutralizing antibodies in baboons. J Virol 1992;66:172–182.

124 Steimer KS, Scandella CJ, Skiles PV, Haigwood NL: Neutralization of divergent HIV-1 isolates by conformation-dependent human antibodies to gp120. Science 1991; 254:105–108.

125 Berkower I, Murphy D, Smith CC, Smith GE: A predominant group-specific neutralizing epitope of human immunodeficiency virus type 1 maps to residues 342 to 511 of the envelope glycoprotein gp120. J Virol 1991;65:5983–5990.

126 Thali M, Olshevsky U, Furman C, Gabuzda D, Posner M, Sodroski J: Characterization of a discontinuous human immunodeficiency virus type 1 gp120 epitope recognized by a broadly reactive neutralizing human monoclonal antibody. J Virol 1991;65:6188–6193.

127 Ho DD, McKeating JA, Li XL, Moudgil T, Daar ES, Sun N-C, Robinson JE: Conformational epitope on gp120 important in CD4 binding and human immunodeficiency virus type 1 neutralization identified by a human monoclonal antibody. J Virol 1991;65:489–493.

128 Laal S, Zolla-Pazner S: Epitopes of HIV-1 glycoproteins recognized by the human immune system; in Norrby (ed): Immunochemistry of AIDS. Chem Immunol. Basel, Karger, 1992, vol 56, pp 91–111.

129 Cordell J, Moore JP, Dean CJ, Klasse PJ, Weiss RA, McKeating JA: Rat monoclonal antibodies to nonoverlapping epitopes of human immunodeficiency virus type 1 gp120 block CD4 binding *in vitro*. Virology 1991;185:72–79.

130 Ho DD, Fung MSC, Cao Y, Li XL, Sun C, Chang TW, Sun N-C: Another discontinuous epitope on glycoprotein gp120 that is important in human immunodeficiency virus type 1 neutralization is identified by a monoclonal antibody. Proc Natl Acad Sci USA 1991;88:8949–8952.

131 Pollard SR, Meier W, Chow P, Rosa JJ, Wiley DC: CD4-binding regions of human immunodeficiency virus envelope glycoprotein gp120 defined by proteolytic digestion. Proc Natl Acad Sci USA 1991;88:11320–11324.

132 Pollard SR, Rosa MD, Rosa JJ, Wiley DC: Truncated variants bind CD4 with high affinity and suggest a minimum CD4 binding region. EMBO J 1992;11:585–591.

133 Pinter A, Honnen WJ, Tilley SA, Bona C, Zaghouani H, Gorny MK, Zolla-Pazner S: Oligomeric structure of gp41, the transmembrane protein of human immunodeficiency virus type 1. J Virol 1989;63:2674–2678.

134 Race EM, Ramsey KM, Lucia HL, Cloyd MW: Human immunodeficiency virus infection elicits early antibody not detected by standard tests: Implications for diagnostics and viral immunology. Virology 1991;184:716–722.

135 Young JAT: HIV and HLA similarity. Nature 1988;333:215.

136 Metlas R, Veljkovic V, Paladini RD, Pongor S: Protein and DNA-sequence homologies between the V3-loop of human immunodeficiency virus type 1 envelope protein gp120 and immunoglobulin variable regions. Biochem Biophys Res Commun 1991; 179:1056–1062.

137 Zaitseva MB, Moshnikov SA, Kozhich AT, Frolova HA, Makarova OD, Pavlikov SP, Sidorovich IG, Brondz BB: Antibodies to MHC class-II peptides are present in HIV-1-positive sera. Scand J Immunol 1992;35:267–273.

138 Cibotti R, Kanellopoulos JM, Cabaniols J-P, Halle-Panenko O, Kosmatopoulos K, Sercarz E, Kourilsky P: Tolerance to a self-protein involves its immunodominant but does not involve its subdominant determinants. Proc Natl Acad Sci USA 1992;89: 416–420.
139 Bost KL, Pascual DW: Antibodies against a peptide sequence within the HIV envelope protein crossreact with human interleukin-2. Immunol Invest 1988;17: 577–586.
140 Horwitz MS, Boyce-Jacino MT, Faras AJ: Novel endogenous sequences related to human immunodeficiency virus type 1. J Virol 1992;66:2170–2179.
141 Pack P, Plückthun A: Miniantibodies: Use of amphipathic helices to produce functional, flexibly linked dimeric F_v fragments with high avidity in *Escherichia coli*. Biochemistry 1992;31:1579–1584.

Dr. A. Robert Neurath, Laboratory of Biochemical Virology,
Lindsley F. Kimball Research Institute, New York Blood Center,
310 East 67th Street, New York, NY 10021 (USA)

Norrby E (ed): Immunochemistry of AIDS.
Chem Immunol. Basel, Karger, 1993, vol 56, pp 61–77

Antigenic and Immunogenic Sites of HIV-2 Glycoproteins[1]

Francesca Chiodi, Ewa Björling, Astrid Samuelsson, Erling Norrby

Department of Virology, Karolinska Institute, Stockholm, Sweden

Evidence of a new human retrovirus serologically distinct from human immunodeficiency virus type 1 (HIV-1) was first reported in 1985 [1] and confirmed by the isolation of HIV-2 in 1986 from two patients with acquired immunodeficiency syndrome (AIDS) who originated from Guinea-Bissau and Cape Verde Islands [2]. In 1987, another HIV-2 strain was isolated in our laboratory from an immunodeficient Gambian woman [3].

Some macaque species, as will be reviewed in the following sections, are susceptible to infection with certain strains of HIV-2 and simian immunodeficiency virus (SIV). The availability of these animal models will contribute to the elucidation of the mechanism of the pathogenesis of immunosuppressive lentiviruses and to disclose immunological responses of protective and non-protective types. HIV-1 and HIV-2 have similar morphological and genomic structures, and they also share many biological properties. A detailed study of the HIV-2 system may, therefore, contribute to a general understanding of protective humoral and cellular immune responses to human lentiviruses.

In the following chapter we will illustrate findings related to the characterization of HIV-2 antigenic and immunogenic sites that stimulate strain- and type-cross-reactive immunity. Corresponding data from preliminary

[1] This work was supported by grants from US Army Medical Research and Development Command (grant No. DAMD17-89-Z-9038) and the Swedish Medical Research Council.

studies of SIV will also be presented. Some epidemiological and biological characteristics of HIV-2 infection will also be reviewed.

Epidemiological Features and Natural History of HIV-2 Infection

HIV-2 is mostly confined to West African countries [4, 5] where the virus seems to have been present for more than 20 years [6]. Several cases have also been reported in Europe and America [5, 7, 8], but most of the infected individuals have resided in or travelled to West Africa.

The initial isolation of the virus from immunodeficient individuals [2, 3] documents that infection with HIV-2, as in the case of HIV-1, can result in the development of AIDS. It is still unclear with what frequency HIV-2 infected individuals progress to full-blown AIDS. A progression rate similar to the one observed for HIV-1 has been reported by several groups [9–11]. Other studies, however, indicate that HIV-2 may have a lower pathogenic potential than HIV-1 and that the time interval from acute infection to overt immunodeficiency may be longer for HIV-2 [6, 12].

Many of the peculiar clinical complications associated with AIDS during HIV-1 infection, such as the appearance of a dementia-like syndrome, are also present in AIDS patients infected with HIV-2 [9]. The patterns of HIV-2 transmission are very similar to what has been observed for HIV-1 and, although mother-to-child transmission of HIV-2 was interpreted to be rare [13], some studies have shown that such a spread of infection may occur [14, 15].

Genomic Organization and Immunological Relatedness of HIV-1 and HIV-2

Analysis of the nucleotide sequences of one HIV-2 isolate, ROD, revealed a genomic organization very similar to that of HIV-1 [16]. In addition to the three structural genes, *gag, pol* and *env*, both viruses have six additional open reading frames coding for proteins with auxiliary functions. One of these proteins, named *vpx* (viral protein X) [17], does not have its counterpart in HIV-1. Conversely, *vpu* (viral protein U) [16] is only found in HIV-1. The two virus serotypes show an amino acid homology of about 60% in both *gag* and *pol* gene products; the envelope proteins are, however, only distantly related, with about 40% identity

[16]. HIV-2 is more closely related to some SIV viruses than HIV-1, and shows a 70–80% overall sequence homology with SIV isolated from macaques (SIV_{mac}) [18] and from sooty mangabey monkeys (SIV_{sm}) [19]. SIV isolated from African green monkeys (SIV_{agm}) [20] and mandrills (SIV_{mnd}) [21] instead appear genetically equidistant from HIV-1 and HIV-2 [22].

The genetic variability observed between different HIV-1 isolates also finds its counterpart in the HIV-2 envelope proteins [23–25]. Variations, however, may not occur with the same high degree observed between HIV-1 strains [26; Stålhandske et al., in preparation].

Because of the 60% homology present in the *gag* and *pol* sequences of HIV-1, HIV-2 and SIV, the capside and nucleocapside proteins of these viruses share antigenic determinants and cross-react in serological tests [27]. Cross-reactivity at the level of the glycoproteins is very low between HIV-1 and HIV-2 strains, whereas HIV-2 and SIV_{mac} are serologically indistinguishable.

In spite of the low degree of homology found in *env* gene products, HIV-1 and HIV-2 use the same receptor, the CD4 molecule, present on the surface of helper T lymphocytes and cells of the macrophage/monocyte lineage [28]. Both HIV-1 and HIV-2 are cytopathic for these cells. HIV-2 isolates, similar to HIV-1 [29], can be divided into groups which show distinct replicative and cytopathic characteristics [25, 30], and the replicative capacity of the virus strains correlates with the clinical severity of the infection in patients [29, 30].

Animal Models for HIV Vaccine Studies

The goals in development and use of a functional vaccine against a micro-organism are to restrict and ideally completely suppress replication upon exposure to the agent. In the practical situation most of the vaccines used in humans do not provide a sterilizing immunity, but replication and spread of the infecting agent are restricted and symptoms of disease are prevented. There is an urge to particularly consider the need for effective containment of lentivirus replication, because of their capacity to integrate in the host genome and to escape immune recognition. During the last few years, the work of several research teams has been focused on the establishment of suitable animal models to evaluate the protective effects of candidate lentivirus vaccines.

Chimpanzees [31] and gibbon apes [32] are the only non-human primates susceptible to HIV-1 infection. Inoculated animals mount an immunological response to the virus, but have not developed any clinical symptoms over a 7-year observation period. The HIV-1 chimpanzee model can, therefore, only be used to study events of sterilizing immunity and, indeed, protection from infection has been achieved in these animals by passive [33a] as well as active [33b, 34] immunization.

Natural infection of African monkeys with SIV has not been associated with the development of immunological disturbances. Experimental SIV infection of Asian macaque species, on the other hand, is considered to be the animal model that best resembles events of AIDS in man. Among the several SIV viruses isolated to date [for review, see ref. 35], only the ones obtained from various macaque species and sooty mangabey monkeys induce an immunodeficiency syndrome in macaques [36–38]. The majority of infected animals show depletion of CD4+ lymphocytes, opportunistic infection, diarrhoea often associated with wasting disease and various degrees of meningo-encephalitis 3 months to 3 years after virus inoculation. Approximately 10% of the SIV inoculated macaques do not show any sign of progression to simian AIDS 2 years after infection. It is not known if resistance to disease progression is conferred by host-genetic factors, reflected in development of competent anti-SIV immunity, or if variations in the virus genome may account for the development of a less pathogenic strain in a few macaques.

Direct extrapolation from SIV models to HIV infection in man should be made with caution. There are distinct antigenic patterns of HIV-1 and SIV. The main sites for neutralizing antibodies appear to be different in the two viruses [see chapter by Javaherian et al. in this volume]. The most attractive model for use in development of HIV vaccines, therefore, would utilize one of the two human lentiviruses to infect and reproduce events of AIDS pathogenesis in primate species that are not too rare and expensive.

Several groups have recently succeeded in infecting macaques with different HIV-2 strains [39–41]. Most of the inoculated animals remain healthy and their CD4 counts are within the normal range. Nevertheless, occasional development of simian AIDS and CD4+ T lymphocyte decrease in HIV-2 inoculated animals have been reported [42, 43]. Since inoculation of monkeys with HIV-2 has been performed only recently, longer observation is needed to evaluate the HIV-2 pathogenic properties in inoculated animals. Because HIV-2 can be used to experimentally infect monkey species that are not endangered, this model offers clear advantages over HIV-1 infection of chimpanzees.

Antibody-Binding Antigenic Sites in the HIV-2 Glycoproteins

Prevention of HIV-2 and SIV_{sm} infection has been achieved in cynomolgus macaques by passive transfer of an anti-SIV_{sm} serum pool with high antibody titres [44]. These results indicate that humoral immunity may play a fundamental role in HIV infection. Consequently, the identification of antibody-binding regions of HIV is a critical issue for studies aimed at vaccine development.

To evaluate the occurrence of dominant linear antigenic sites in the envelope glycoproteins of HIV-2, we studied the reactivity of sera obtained from HIV-2 and HIV-1 infected humans and SIV_{sm} inoculated monkeys to peptides representing different regions of the HIV-2 strain SBL-6669 [45]. Four major antigenic sites were identified, two in the gp125 external protein and two in the transmembrane glycoprotein (fig. 1). The dominant antigenic region identified in the HIV-2 V3 region consisted of two distinct components, one of strain-specific nature, corresponding to the central part of the V3 loop, and one cross-reactive site in the carboxy-terminal part. Also the carboxy-terminal part of the gp125 (Glu^{472}–Arg^{493}) contained cross-reactive determinants. Two additional type-specific antigenic sites were identified in the amino-terminal part of gp36 protein between amino acids Ala^{573}–Cys^{595} and Glu^{634}–Lys^{649}. As previously reported [47, 48], the use of the peptide representing the region Ala^{573}–Cys^{595} in serological tests allows the recognition of all antibody-positive sera; this highly antigenic region has been mapped to the amino-terminal part of the transmembrane protein of all known human and simian lentiviruses [reviewed in ref. 49]. The second antigenic site of the transmembrane protein is unique for the HIV-2 virus and recognized only by HIV-2 and SIV sera.

The chemistry of site-directed serology based on the highly antigenic region in the amino-terminal part of the transmembrane protein has been evaluated by use of both simian and human post-infection sera [50] and of antipeptide monoclonal antibodies (MoAbs) [51]. By use of deletion and substitution sets of peptides, a single dominant maximally 7-amino-acid-long antigenic site was demonstrated at Trp^{596}–Gln^{602} (SIV_{mac}) for HIV-2 and SIV antibodies. A similar analysis of the antigenic specificity of HIV-1 sera showed dependence on the corresponding amino acids Gly^{597}–Leu^{602}, but in addition a role in a separate site of Lys^{598} and Asp^{599}. Dissection of the immunogenic properties of the SIV_{mac} peptide Ala^{582}–Cys^{604} by use of 34 murine MoAbs demonstrated the occurrence of five different antigenic sites. Within some sites the antibodies could be subgrouped to show a progres-

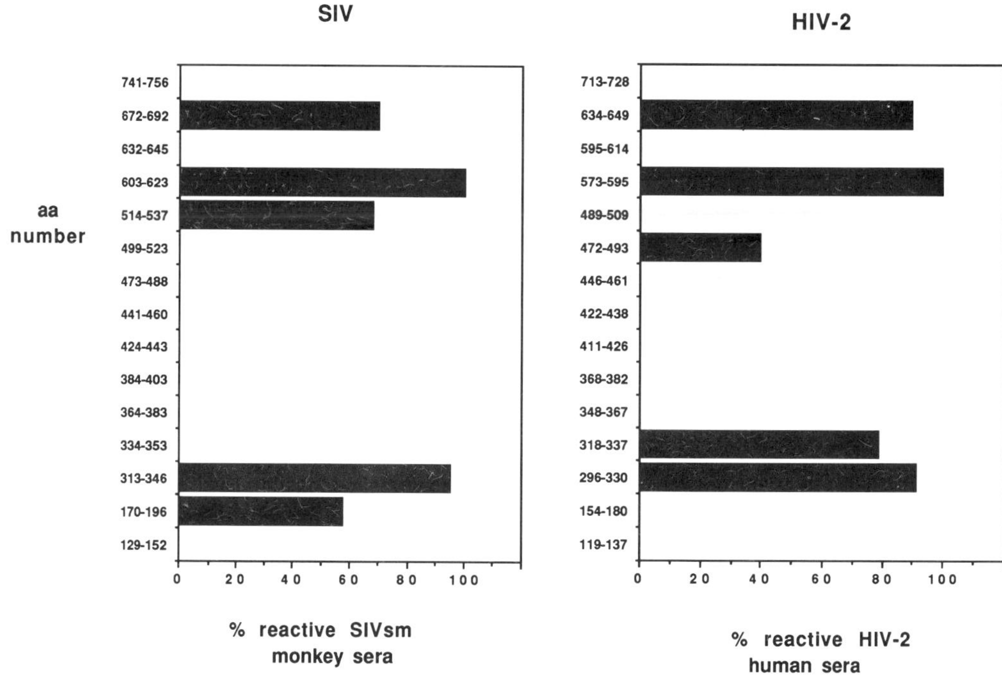

Fig. 1. Comparison of antibody-binding regions of HIV-2 and SIV_{sm}. The peptide amino acid sequences were derived from the published sequences of the ISY molecular clone of HIV-2 strain SBL-6669 and from the SIV_{smmH4} of SIV_{sm} [46]. The numbers on the vertical axis indicate the amino acid position of the peptides in the envelope proteins. HIV-2 and SIV_{sm} peptides are aligned to represent analogous regions.

sively more narrow epitopic dependence on amino acids in the central part of the site. Antipeptide MoAbs reacting with the site Trp^{596}–Gln^{602} effectively blocked the capacity of the peptide to react with human post-infection HIV-2 antibodies. No similar blocking was seen with MoAbs to a neighbouring site, Leu^{587}–Glu^{590}. An intertype cross-reacting site was identified in the amino-terminal region of the corresponding HIV-2 and HIV-1 peptides, Ala^{582}–Glu^{588}/Lys^{588}. However, this site was not accessible in the intact transmembrane protein. The dominating antigenic site of the transmembrane protein, represented by SIV peptides, was shown by nuclear magnetic resonance analysis [52] to have a propensity to fold in aqueous solution.

The presence of highly immunogenic domains in the extracellular proteins of HIV-2 has also been demonstrated by Western blot analysis using bacterially expressed contiguous segments representing the HIV-2 envelope products. By this approach, Huang et al. [53] identified two major antigenic sites comprising amino acids 224–380 and 524–763 of the ST strain of HIV-2, likely to correspond to the regions identified by us in the corresponding V3 loop and amino-terminal part of the transmembrane protein. Over 95% of HIV-2 positive sera from Senegal reacted with these protein segments.

Goudsmit et al. [54], by using the antibody-reactive peptide scanning method, defined eight distinct antibody-binding sites in the transmembrane protein gp36 of the HIV-2 strain ROD in addition to the highly immunodominant site present at the amino terminus of gp36. A minority of the HIV-2 antibody positive sera used in the study reacted to these eight regions [54] and only five of the antibody-binding regions were predicted to be exposed at the surface of the virus or infected cell membrane.

The location of antigenic sites in the envelope proteins of SIV_{sm} [55] is similar to that found in HIV-1 and HIV-2 systems (fig. 1), emphasizing the utility of the HIV-2 and SIV monkey model for vaccine studies. A unique antigenic region was identified in the V2 region of SIV_{sm} (amino acids 170–196); the corresponding peptide recognizes approximately 60% of the infected SIV animals but none of the HIV-2 antibody positive sera from humans or experimentally inoculated animals. Lack of antibodies to the SIV V2 region appears to be associated with disease progression in the infected macaques.

Strain- and Type-Cross-Reactive Neutralization of HIV-2 Antibody Positive Sera

The characterization of neutralizing responses in the serum of HIV-1 infected individuals has been the focus of numerous studies, as reviewed in other chapters of this volume. For several reasons, including the restricted number of HIV-2 infected persons and the occurrence of the epidemic only in West African countries, the serum neutralization patterns of HIV-2 sera are not as well defined as in HIV-1.

Neutralizing antibodies to the HIV-2 virus are present in the serum of infected individuals. Weiss et al. [56] examined a restricted panel of HIV-2 and HIV-1 antibody sera by VSV (HIV) pseudotype neutralization against

HIV-1 and HIV-2 isolates. The HIV-2 West African sera could cross-neutralize both virus serotypes, whereas HIV-1 sera inhibited virus replication only in a type-specific fashion. Due to the presence of broadly cross-protective HIV-2 antibodies, the authors suggested that HIV-2 antigens should be included in a hypothetical AIDS vaccine.

Different results were obtained in a study conducted on a larger material by Böttiger et al. [57]. When examining 57 HIV-1 and 43 HIV-2 antibody positive sera for neutralizing capacity against two HIV-1 isolates (IIIB and RF) and two HIV-2 isolates (SBL-6669 and -K135) they found the presence of cross-reacting, type-specific or strain-specific neutralizing antibodies in both HIV-1 and HIV-2 sera. Moreover, cross-neutralizing titres were lower than type-specific neutralizing titres in the same serum specimens. No differences could be seen in neutralizing titres between different clinical stages in HIV-1 and HIV-2 infected individuals.

Neutralizing antibodies to HIV-2 inoculum develop in cynomolgus monkeys infected with HIV-2 strains [58]. Animals with a post-HIV-2 infection immunity were shown to be protected from disease development upon challenge with SIV_{sm} virus, possibly inferring a role for cross-neutralizing antibodies in disease attenuation.

The presence of cross-reacting neutralizing antibodies between HIV-1 and HIV-2 does indicate that the targets for neutralizing antibodies are distinct from those envelope antigens which bind to antibodies mediating cellular cytotoxicity (ADCC). In a study conducted in our laboratory, Ljunggren et al. [59] found no cross-reactive ADCC activity between HIV-1 and HIV-2 strains.

Target Regions for Neutralizing Antibodies in the HIV-2 Glycoproteins

A number of linear and conformational target regions for neutralizing antibodies have been identified in the envelope glycoproteins of HIV-1, whereas the immunoprotective epitopes of HIV-2 glycoproteins remain to be mapped. One approach for the identification of linear immunogenic epitopes is to assay the neutralizing capacity of animal immune sera raised against synthetic peptides corresponding to defined protein regions.

By inoculating guinea pigs with 25 peptides from the envelope proteins of the HIV-2 isolate SBL-6669, we could identify five different target regions for neutralization [60], three sites in gp125 and two sites in gp36. The most active immunogenic site was located in the central and carboxy-terminal part of the HIV-1 V3-homologous region. Part of the V1 region (amino acids 119–137) and a region at the carboxy-terminal end of the gp125 were also found to be capable of inducing neutralizing antibodies. Two peptides representing regions of gp36 elicited neutralizing antibodies in the immunized guinea pigs. The first region (amino acids 595–614) is located carboxy-terminally to the highly antigenic sites, which allows type-specific serology [48], whereas the second immunogenic site of the transmembrane protein (amino acids 714–729) is predicted in analogy with HIV-1, to be cytoplasmic [61].

The role of HIV-1 V3 in inducing neutralizing antibodies has been highlighted by several studies [for review, see chapters by Neurath and Laal and Zolla-Pazner in this volume]. Protein prediction models, however, indicate that the structural characteristics of the V3 region of HIV-1 and HIV-2 show considerable divergence. We have sequenced the V3-homologous region of 12 HIV-2 isolates from Guinea-Bissau [Stålhandske et al., in preparation] by polymerase chain reaction and cloning of the amplified fragments. The comparison of our results with the published sequences of other HIV-2 isolates [46] revealed a lower ratio of variability in comparison to HIV-1 V3 [62].

The inter-isolate variability observed in the principal neutralizing domain of HIV-1 has been ascribed to immune selection. Studies aimed at examining genetic variation of molecular clones of SIV proviral DNA in experimentally infected macaques [63, 64] have shown that the SIV region corresponding to HIV-1 V3 may not be subjected to the same selective pressure and, thus, it remains relatively conserved in all animals during progression of infection. These studies indicate that this region might not be the main target for neutralizing antibodies in the SIV system.

The observations reported by Javaherian et al. [65] reinforce the latter hypothesis. Synthetic peptides representing the SIV region corresponding to HIV-1 V3 and reduced gp110 SIV glycoprotein did not inhibit neutralizing antibodies in sera from SIV-infected macaques and elicited very low neutralizing antibody titers when inoculated in the animals. The use of native gp110 in blocking experiments, on the contrary, resulted in complete inhibition of neutralizing activity by sera of inoculated animals. Accordingly, the authors suggested that the principal neutralizing domain of SIV might be represented by a conformational epitope. We conducted similar experiments and succeeded in the attempt of blocking neutralizing activity of HIV-2 human sera with peptides derived from the HIV-2 V3 sequence [66]. This result indicates that a V3

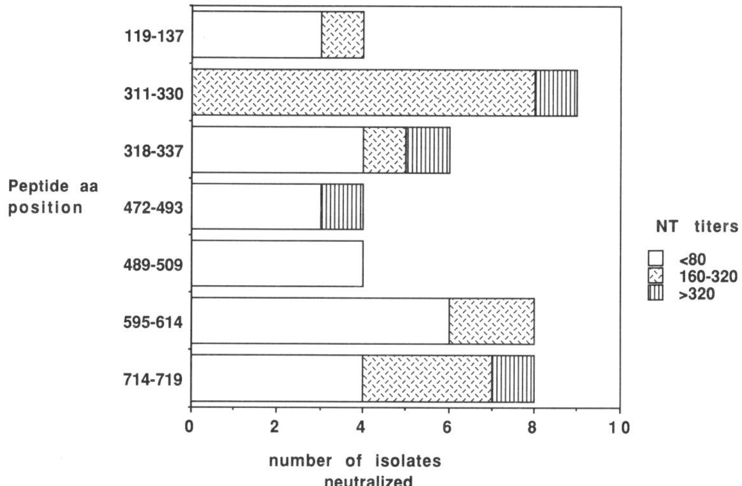

Fig. 2. Cross-neutralization activity of anti-HIV-2 envelope peptide guinea pig sera with 10 strains of HIV-2. The amino acid positions indicate the peptides used to immunize different guinea pigs.

linear domain(s) represent a dominant target for neutralizing antibodies in the HIV-2 system.

Immunization of animals with linear peptides is likely to result in the efficient presentation of protective immunogenic sites that may be poorly represented in whole antigen preparations. We assayed the capacity of guinea pig sera raised against HIV-2 envelope peptides [60] to mediate cross-neutralization of several West African isolates [66]. Ten isolates from which V3 sequences are available were used in the study, including the LAV-2 strain. The results of the neutralization experiments are illustrated in figure 2. Antibodies to the HIV-2 V3 region (amino acids 311–330; peptide sequence SGRRFHSQKIINKKPRQAWC; the V3 amino acid sequences of some HIV-2 isolates are shown in figure 3) cross-neutralized nine of the ten isolates included in the study, with titres ranging between 160 and 640. Antibodies to peptides representing regions of the transmembrane protein (amino acids 595–614 and 714–729) mediated cross-neutralization of a large number of HIV-2 isolates, but with low titres (fig. 2). Both regions are relatively conserved among the different HIV-2 isolates (fig. 3), thus explaining cross-neutralizing of anti-HIV-2 monospecific sera.

```
HIV-2  ISY   CRRPENKTVVPITLMSGRRFHS..QKIINKKPRQAWC
       ROD    K  G  I KQ M    HV   HY P .  R
       NIHZ   K  G     L  F   FK      PV
       ST     K  G            LV      P . RR
       BEN    K  G     L      LV      P . TR
       D194   K  G                   RPVY    G
       GH1    K  G            LV      P . TR
A      D205   K  G       RTV  LL      P .  R
```

```
HIV-2  ISY   CHTTVPWVNDTLTPEWNNMT
       ROD             S A D D
       NIHZ     S          D
       ST                  D
       BEN             S S D K
       D194            S   D
       GH1             S S D
B      D205         P E    N
```

```
HIV-2  ISY   HIHKDWEQPDREETEE
       ROD        RG  AN
       NIHZ       Q   A
       ST         R   A
       BEN        RG  AN G
       D194     T RG  AN
       GH1      T RG  AN G
C      D205  P R  RG  AN
```

Fig. 3. Comparison of HIV-2 amino acid sequences corresponding to three targets for cross-neutralization. The published sequences of HIV-2 isolates [46] are compared to the ISY molecular clone of the SBL-6669 strain, which sequence was used to synthesize the peptides used in the immunization procedure.

Concluding Remarks

Antibody reactions of SIV_{sm}-infected macaque sera against selected SIV_{sm} envelope peptides are very similar to those of HIV-2- and HIV-1-infected human sera to the corresponding linear antigenic sites. These findings reinforce the existence of immunological parallelism between human and simian lentiviruses. HIV-2-induced immunity in inoculated macaques has been shown to protect against SIV-associated disease [58], indicating that, indeed, the two viruses share group-specific protective immunity. Whether or not linear domains play a role in eliciting SIV-neutralizing responses remains to be defined.

Linear envelope regions of HIV-2 can mediate antigenic and immunogenic responses [45, 60]. In a previous study of human sera, Böttiger et al. [67] found that the occurrence of an intertype cross-reacting V3-region-specific activity correlated with the presence of cross-neutralizing activity in sera. More recently, it was found [Putkonen et al., in preparation] that primary immunization of cynomolgus monkeys with whole HIV-2 virus antigen, followed by booster with synthetic peptides representing the HIV-2 V3-homologous loop, appears to protect against infection (three of four animals) after challenge with homologous HIV-2 virus (SBL-6669). These findings, along with the blocking of human neutralizing antibodies by V3 peptides [66], indicate that the V3 region may be an important neutralizing site not only in HIV-1 but also in HIV-2.

Our current approach to define HIV-2 and SIV protective immunogenic sites is to use whole virus antigen for primary immunization and then to evaluate the usefulness of boostering with different combinations of peptides. The challenge with SIV and HIV-2 homologous virus and heterologous virus strains will indicate which regions play a role in determining protection from infection and disease.

References

1 Barin F, Denis F, Allan JS, M'Boup S, Kanki P, Lee TH, Essex M: Serological evidence for virus related to simian T-lymphotropic virus III in residents of West Africa. Lancet 1985;ii:1387–1389.

2 Clavel F, Guétard D, Brun-Vézinet F, Chamaret S, Rey M-A, Santos-Ferreira MO, Laurent AG, Dauguet C, Katlama C, Rouzloux C, Klatzmann D, Champalimaud JL, Montagnier L: Isolation of a new human retrovirus from West-African patients with AIDS. Science 1986;233:343–346.

3 Albert J, Bredberg U, Chiodi F, Böttiger B, Fenyö EM, Norrby E, Biberfeld G: A new human retrovirus isolate of West African origin (SBL-6669) and its relationship to HTLV-IV, LAV-II and HTLV-IIIB. AIDS Res Hum Retroviruses 1987;3:3–10.

4 Kanki PJ, M'Boup S, Ricard D, Barin F, Denis F, Boye C, Sangare L, Travers K, Albaum M, Marlink R, Romet-Lemonne J-L, Essex M: Human T-lymphotropic virus type 4 and the human immunodeficiency virus in West Africa. Science 1987; 236:827–831.

5 De Cock KM, Brun-Vezinet F: Epidemiology of HIV-2 infection. AIDS 1989;3 (suppl 1):S89–S95.

6 Bryceson A, Tomkins A, Ridley D, Warhurst D, Goldstone A, Bayliss G, Toswill J, Parry J: HIV-2 associated AIDS in the 1970s. Lancet 1988;ii:221.

7 Centers for Disease Control: Update HIV-2 infection – United States. MMWR 1989; 38:572–580.
8 Veronesi R, Mazza CC, Santos Ferreira MO, Lourenco MH: HIV-2 in Brazil. Lancet 1987;ii:402.
9 Clavel F, Mansinho K, Chamaret S, Guetard D, Favier V, Nina J, Santos-Ferreira M-D, Champalimaud J-L, Montagnier L: Human immunodeficiency virus type 2 infection associated with AIDS in West Africa. N Engl J Med 1987;316:1180–1185.
10 Brun-Vezinet F, Rey MA, Katlama C, Roulot D, Lenoble L, Alizon M, Madjar JJ, Rey MA, Girard PM, Yeni P, Clavel F, Gadelle S, Harzic M: Lymphoadenopathy-associated virus type 2 in AIDS and AIDS-related complex. Lancet 1987;i:128–132.
11 Nauclér A, Albino P, Da Silva AP, Andreasson PÅ, Andersson S, Biberfeld G: HIV-2 infection in hospitalized patients in Bissau, Guinea-Bissau. AIDS 1991;5:301–304.
12 Ancelle R, Bletry O, Baglin AC, Brun-Vezinet F, Rey MA, Godeam P: Long incubation time for HIV-2 infection. Lancet 1987;i:688–689.
13 Poulsen AG, Kvinesdal B, Aaby P, et al: Prevalence of and mortality from human immunodeficiency virus type 2 in Bissau, West Africa. Lancet 1989;i:827–830.
14 Gnaore E, De Cock KM, Gayle H, et al: Prevalence of and morbidity from HIV type 2 in Guinea-Bissau, West Africa. Lancet 1989;ii:513.
15 Morgan G, Wilkins HA, Pepin J, Jobe O, Brewster D, Whittle H: AIDS following mother-to-child transmission of HIV-2. AIDS 1990;4:879–882.
16 Guyader M, Emerman M, Sonigo P, Clavel F, Montagnier L, Alizon M: Genome organization and transactivation of the human immunodeficiency virus type 2. Nature 1987;326:662–669.
17 Henderson LE, Sowder RC, Copeland TD, Benveniste RE, Oroszlan S: Isolation and characteristics of a novel protein (X-ORF product) from SIV and HIV-2. Science 1988;241:199–202.
18 Daniel MD, Letvin NL, King NW, Kannagi M, Sehgal PK, Hunt RD, Kanki PJ, Essex M, Desrosiers RC: Isolation of the T-cell tropic HTLV-III-like retrovirus from macaques. Science 1985;228:1201–1204.
19 Murphy-Corb M, Martin LN, Rangan SR, Baskin GB, Gormus BJ, Wolf RH, Abe Andes W, West M, Montelaro RC: Isolation of an HTLV-III related retrovirus from macaques with simian AIDS and its possible origin in asymptomatic mangabeys. Nature 1986;321:435–437.
20 Daniel MD, Ythirajulu M, Naidu M, Durda PJ, Schmidt DK, Troup CD, Silva DP, MacKey JJ, Kestler HW III, Sehgal PK, King NW, Ohta Y, Haygami M, Desrosiers RC: Simian immunodeficiency virus from African green monkeys. J Virol 1988; 62:4123–4128.
21 Tsujimoto H, Cooper RW, Komada T, Fukasawa M, Miura T, Ohta Y, Ishikawa K-I, Nakai M, Frost E, Roelants GE, Roffi J, Hayami M: Isolation and characterization of simian immunodeficiency virus from mandrills in Africa and its relationship to other human and simian immunodeficiency viruses. J Virol 1988;62:4044–4050.
22 Fukasawa M, Miura T, Hasegawa A, Morikawa S, Tsujimoto H, Mikl K, Kitamura T, Hayami M: Sequences of simian immunodeficiency virus from African green monkey, a new member of the HIV/SIV group. Nature 1988;333:457–461.

23 Zagury JF, Franchini G, Reitz M, Collalti E, Starcich B, Hall L, Fargnoli K, Jagodzinski L, Guo HG, Laure F, Arya SK, Josephs S, Zagury Z, Wong-Staal F, Gallo RC: Genetic variability between HIV-2 isolates is comparable to the variability among HIV type 1. Proc Natl Acad Sci USA 1988;85:5941–5945.
24 Dietrich U, Adamski M, Kreutz R, Seipp A, Kühnel H, Rübsamen-Waigmann H: A highly divergent HIV-2 related isolate. Nature 1989;342:948–950.
25 Schulz TF, Whitby D, Hoad JG, Corrah T, Whittle H, Weiss RA: Biological and molecular variability of human immunodeficiency type 2 isolates from The Gambia. J Virol 1990;64:5177–5182.
26 Gardner MB, Luciw PA: Simian immunodeficiency viruses and their relationship to the human immunodeficiency virus. AIDS 1988;2(suppl 1):S3–S10.
27 Biberfeld G, Böttiger B, Bredberg-Råden U, Putkonen P, Starup C, Håkansson C: Findings in four HTLV-IV seropositive women from West Africa. Lancet 1986;ii: 1330–1331.
28 Sattentau QJ, Clapham PR, Weiss R, Beverley PCL, Montagnier L, Alhalabi MF, Gluckmann JC, Klatzmann D: The human and simian immunodeficiency virus HIV-1, HIV-2 and SIV interact with similar epitopes on their cellular receptor, the CD4 molecule. AIDS 1988;2:101–105.
29 Åsjö B, Morfeldt-Månsson L, Albert J, Biberfeld G, Karlsson A, Lidman K, Fenyö EM: Replicative capacity of human immunodeficiency virus from patients with varying severity of HIV infection. Lancet 1986;ii:660–662.
30 Albert J, Nauclér A, Böttiger B, Broliden P-A, Albino P, Quattara SA, Björkegren C, Valentin A, Biberfeld G, Fenyö EM: Replicative capacity of HIV-2, like HIV-1, correlates with severity of immunodeficiency. AIDS 1990;4:291–295.
31 Alter HJ, Eichberg JW, Masur H, Saxinger WC, Gallo RC, Macher AM, Lane HC, Fauci AS: Transmission of HTLV-III infection from human plasma to chimpanzees. An animal model for AIDS. Science 1984;226:549–552.
32 Lusso P, Markham PD, Ranki A, Earl P, Moss B, Dorner F, Gallo RC, Krohm KJE: Cell-mediated immune response toward viral envelope and core antigens in gibbon apes *(Hylobates lar)* chronically infected with human immunodeficiency virus. J Immunol 1988;141:2467–2473.
33a Emini EA, Schleif WA, Nunberg JH, Conley AJ, Eda Y, Tokiyoshi S, Putney SD, Matsushita S, Cobb KE, Jett CM, Eichberg JW, Murthy KK: Prevention of HIV-1 infection in chimpanzees by gp 120 V3 domain-specific monoclonal antibody. Nature 1992;355:728–730.
33b Berman PW, Gregory TJ, Riddle L, Nakamura GR, Champe MA, Porter JP, Wurm FM, Hershberg RD, Cobb EK, Eichberg JW: Protection of chimpanzees from infection by HIV-1 after vaccination with recombinant glycoprotein gp120 but not gp160. Nature 1990;345:622–624.
34 Girard M, Kiney MP, Pinter A, Barre-Sinoussi F, Nara P, Kolbe H, Kusumi K, Chaput A, Reinhart T, Muchmore E, Ronco J, Kaczorek M, Gomard E, Gluckman J-C: Immunization of chimpanzees confers protection against challenge with human immunodeficiency virus. Proc Natl Acad Sci USA 1991;88:542–546.
35 Desrosiers RC: The simian immunodeficiency viruses. Annu Rev Immunol 1990;8: 557–578.
36 Putkonen P, Warstedt K, Thorstensson R, Benthin R, Albert J, Lundgren B, Öberg B, Norrby E, Biberfeld G: Experimental infection of cynomolgus monkeys *(Macaca fascicularis)* with simian immunodeficiency virus (SIVsm). J AIDS 1989;2:359–365.

37 Daniel MD, Letvin NL, Seghal PK, Hunsmann G, Schmidt D, King NW, Desrosiers R: Long-term persistent infection of Macaque monkeys with the simian immunodeficiency virus. J Gen Virol 1987;68:3183–3189.

38 Benveniste RE, Morton WR, Clark EA, Tsai CC, Ochs HD, Ward JM, Kuller L, Knott WB, Hill RW, Gale MJ, Thoeless ME: Inoculation of baboons and macaques with simian immunodeficiency virus/Mne, a primate lentivirus closely related to human immunodeficiency virus type 2. J Virol 1988;62:2091–2101.

39 Letvin NL, Daniel PK, Sehgal J, Yetz JM, Solomon KR, Kannagi M, Schmidt DK, Silva DP, Montagnier L, Desrosiers RC: Infection of baboons with human immunodeficiency virus-2 (HIV-2). J Inf Dis 1987;156:406–407.

40 Franchini G, Markham P, Gard E, Fargnolli K, Keubaruwa S, Jagodzinski L, Robert-Guroff M, Lusso P, Ford G, Wong-Staal F, Gallo RC: Persistent infection of rhesus macaques with a molecular clone of human immunodeficiency virus type 2: Evidence of minimal genetic drift and low pathogenetic effects. J Virol 1990;64: 4462–4467.

41 Putkonen P, Böttiger B, Warstedt K, Thorstensson R, Albert J, Biberfeld G: Experimental infection of cynomolgus monkeys *(Macaca fascicularis)* with HIV-2. J AIDS 1989;2:366–367.

42 Dormont D, Livartowsky J, Chamaret S, Guetard D, Henin D, Levagueresse R, van der Moortelle PF, Larke B, Gourmelon P, Vazeux R, Metivier H, Flageat J, Court L, Hauw JJ, Montagnier L: HIV-2 in rhesus monkeys: Serological virological and clinical results. Intervirology 1989;30(suppl 1):59–65.

43 Dortmont D, Livartowski J, Vogt G: Second in vivo passage of HIV-2 in rhesus monkeys; in Schallekens H, Horzinek MC (eds): Animal Model in AIDS. Amsterdam, Elsevier, 1990, pp 63–71.

44 Putkonen P, Thorstensson R, Ghavamzadeh L, Albert J, Hild K, Biberfeld G, Norrby E: Prevention of HIV-2 and SIVsm infection by passive immunization in cynomolgus monkeys. Nature 1991;352:436–437.

45 Norrby E, Putkonen P, Böttiger B, Utter G, Biberfeld G: Comparison of linear antigenic sites in the envelope proteins of human immunodeficiency virus (HIV) type 2 and type 1. AIDS Res Hum Retroviruses 1991;7:279–285.

46 Myers G, Korber B, Berzofsky JA, Smith RF, Pavlakis GN (eds): Database Human Retroviruses AIDS. Las Alamos, National Laboratory, 1991.

47 Gnann JW Jr, Schwimmbeck PL, Nelson JA, Truax AB, Oldstone MB: Diagnosis of AIDS by using a 12 amino acid peptide representing an immunodominant epitope of the human immunodeficiency virus. J Infect Dis 1987;156:261–267.

48 Norrby E, Biberfeld G, Chiodi F, von Gegerfeldt A, Naucler A, Parks E, Lerner R: Discrimination between antibodies to HIV and to related retroviruses using site-directed serology. Nature 1987;329:248–250.

49 Norrby E: Human immunodeficiency virus antibody responses determined by site-directed serology. Intervirology 1990;31:315–326.

50 Norrby E, Parks DE, Utter G, Houghten RA, Lerner RA: Immunochemistry of the dominating antigenic region Ala^{582} to Cys^{604} in the transmembranous protein of simian and human immunodeficiency virus. J Immunol 1989;143:3602–3608.

51 Norrby E, Biberfeld G, Johnson PR, Parks DE, Houghten RA, Lerner RA: The chemistry of site-directed serology for HIV infections. AIDS Res Hum Retroviruses 1989;5:487–493.

52 Dyson HJ, Norrby E, Hoey K, Parks DE, Lerner RA, Wright PE: Immunogenic

peptides corresponding to the dominant antigenic region Ala^{597} to Cys^{619} in the transmembrane protein of simian immunodeficiency virus have a high folding propensity. Biochemistry 1992;31:1458–1463.

53 Huang ML, Essex M, Lee T-H: Localization of immunogenic domains in the human immunodeficiency virus type 2 envelope. J Virol 1991;65:5073–5079.

54 Goudsmit J, Meloen RH, Brasseur R, Barin F: Human B-cell epitopes of HIV-2 transmembrane protein are similarly spaced as in HIV-1. J AIDS 1989;2:297–302.

55 Samuelsson A, Björling E, Putkonen P, Utter G, Chiodi F, Norrby E: Identification of four antibody-binding sites in the envelope proteins of simian immunodeficiency virus (SIVsm). Submitted.

56 Weiss RA, Clapham PR, Weber JN, Whitby D, Tedder RS, O'Connor T, Chamaret S, Montagnier L: HIV-2 antisera cross-neutralize HIV-1. AIDS 1988;2:95–100.

57 Böttiger B, Karlsson A, Naucler A, Andreasson PÅ, Mendes Costa C, Biberfeld G: Cross-neutralizing antibodies against HIV-1 (HTLV-IIIB and HTLV-IIIRF) and HIV-2 (SBL-6669 and a new isolate SBL-K135). AIDS Res Hum Retroviruses 1989; 5:511–519.

58 Putkonen P, Thorstensson R, Albert J, Hild K, Norrby E, Biberfeld P, Biberfeld G: Infection of cynomolgus monkeys with HIV-2 protects against pathogenic consequences of a subsequent simian immunodeficiency virus infection. AIDS 1990;4: 783–789.

59 Ljunggren K, Chiodi F, Biberfeld G, Norrby E, Jondal M, Fenyö EM: Lack of cross-reaction in antibody-dependent cellular cytotoxicity between human immunodeficiency virus (HIV) and HIV-related West African strains. J Immunol 1988;140:602–605.

60 Björling E, Broliden K, Bernardi D, Utter G, Thorstensson R, Chiodi F, Norrby E: Hyperimmune antisera against synthetic peptides representing the glycoprotein of human immunodeficiency virus type 2 can mediate neutralization and antibody-dependent cytotoxic activity. Proc Natl Acad Sci USA 1991;88:6082–6086.

61 Chahn T, Dreesman G, Kanda P, Linette G, Sparrow J, Ho D, Kennedy R: Induction of anti-HIV neutralizing antibodies by synthetic peptides. EMBO J 1986;11:3065–3071.

62 La Rosa GJ, Davide JP, Weinhold K, Waterbury J, Profy A, Lewis J, Langlois A, Dreesman G, Boswell R, Shadduck P, Holley L, Karplus M, Bolognesi D, Matthews T, Emini E, Putney S: Conserved sequence and structural elements in the HIV-1 principal neutralizing determinant. Science 1990;249:932–935.

63 Johnson RR, Hamm TE, Goldstein S, Kitov S, Hirsch VM: The genetic fate of molecularly cloned simian immunodeficiency virus in experimentally infected macaques. Virology 1991;185:217–228.

64 Burns DPW, Desrosiers RC: Selection of genetic variants of simian immunodeficiency virus in persistently infected rhesus monkeys. J Virol 1991;65:1843–1854.

65 Javaherian K, Langlois AJ, Schmidt S, Kaufmann M, Cates N, Langedijk JPM, Meloen RH, Desrosiers RC, Burns DPW, Bolognesi DP, La Rosa GJ, Putney SD: The principal neutralization determinant of simian immunodeficiency virus differs from that of human immunodeficiency virus type 1. Proc Natl Acad Sci USA 1992; 89:1418–1422.

66 Björling E, Chiodi F, Norrby E: Targets for cross-neutralization in the human immunodeficiency virus type 2 (HIV-2) glycoproteins. Submitted.

67 Böttiger B, Karlsson A, Andreasson P-Å, Nauclér A, Costa CM, Norrby E, Biberfeld G: Envelope cross-reactivity between HIV-1 and HIV-2 detected by different serological methods: Correlation between cross neutralization and reactivity against the main neutralizing site. J Virol 1990;64:3492–3499.

Francesca Chiodi, PhD, Department of Virology, Karolinska Institute,
c/o SBL, Lundagatan 2, S–105 21 Stockholm (Sweden)

Norrby E (ed): Immunochemistry of AIDS.
Chem Immunol. Basel, Karger, 1993, vol 56, pp 78–90

SIV Neutralization Epitopes

Kashi Javaherian[a], *Alphonse J. Langlois*[b], *Gregory J. LaRosa*[a]

[a] Repligen Corporation, Cambridge, Mass.;
[b] Duke University Medical School, Durham, N.C., USA

Developing a vaccine against the acquired immunodeficiency syndrome (AIDS) has been the main goal of a large number of investigators throughout the world. Although attempts have been made to make use of the inactivated human immunodeficiency virus for a vaccine, due to the risks associated with this method it has not been considered a plausible option. Therefore, attention has been mainly focused on the subunit proteins, in particular the HIV envelope. This protein contains a number of important immunological determinants which make it an attractive vaccine candidate. It has been shown to carry the principal neutralization determinants (PND) which elicit neutralizing antibodies [1–3], an important form of protective immunity [4]. Employing a number of recombinant subunits corresponding to different parts of the envelope protein followed by peptide mapping, we concluded that the PND was located in a dusulfide closed loop in the hypervariable V3 region of the envelope [1, 5, 6]. Although this region of the envelope elicits isolate-specific neutralizing antibody, we discovered that there was a small sequence of conserved amino acids within the PND which was capable of eliciting a broad neutralization response [7, 8]. More recently, it has been demonstrated that the PND bears sequences serving as cytotoxic T lymphocyte (CTL) determinants [9]. By modulating few amino acids in the CTL domain, the authors were able to induce broadly cross-reactive cytotoxic T cells. The combination of humoral and CTL responses can be powerful ingredients for a vaccine formulation. A cocktail of envelope proteins corresponding to the neutralization and CTL determinants of the predominant viral isolates [8] may be considered as an attractive vaccine candidate. In addition to the above considerations, the envelope protein in its native form binds the CD4 receptor. Antibodies to the CD4-binding domain of the external envelope gp120 neutralize diverse viral isolates [10–18]. However, this form of neutralizing antibody, which is conformation-dependent, comprises a small percentage of the total neutralizing antibody response to gp120 [11]. One is inclined to conclude that employing the envelope protein in its native form has the additional advantage of eliciting these broadly neutralizing antibodies in addition to those associated with PND, which were shown to be independent of conformation [1]. In this connection, it should be pointed out that the neutralizing antibodies obtained from chimpanzees immunized with a single native gp120 mapped to the PND of the envelope and a correlation was observed between neutralizing antibody and protection against HIV [19].

One major problem with HIV research is the lack of suitable animal models for testing. Chimpanzees as the closest species to humans have been used for this purpose. However, they are expensive and although susceptible to infection by HIV they do not develop disease. For these and other reasons, attention has turned to the simian immunodeficiency virus (SIV) which is similar to HIV but infects monkeys. They develop an AIDS-like disease once exposed to pathogenic forms of the virus. Two types of successful protection experiments have been performed with macaques against SIV. In one method, animals were immunized with the inactivated virus followed by live viral challenge [for a review, see ref. 20]. In a more recent trial, protective immunity was observed by immunizing macaques with recombinant vaccinia virus expressing SIV gp160 and then boosted with recombinant SIV envelope protein [21]. We have been interested in SIV primarily to study the role of neutralizing antibodies in protection against infection. Toward this goal, we have attempted to define the PND of SIV. Our hypothesis was based on the premise that, if PNDs of SIV and HIV were similar, the feasibility of the envelope as a vaccine candidate can be directly tested in monkeys. We report here that linear peptides corresponding to the V3 region of the SIV envelope protein, in contrast to HIV-1, do not elicit neutralizing antibody [22]. Furthermore, the native structure of the envelope is essential for neutralizing antibody in SIV.

Some Structural Characteristics of the SIV Envelope

Since the SIV envelope protein plays an important role in the following discussions, it would be appropriate to discuss some of the properties of this protein in more detail. SIV-Mac251 gp110 used in our studies is a 505-amino-acid protein corresponding to the outer membrane portion of the SIV

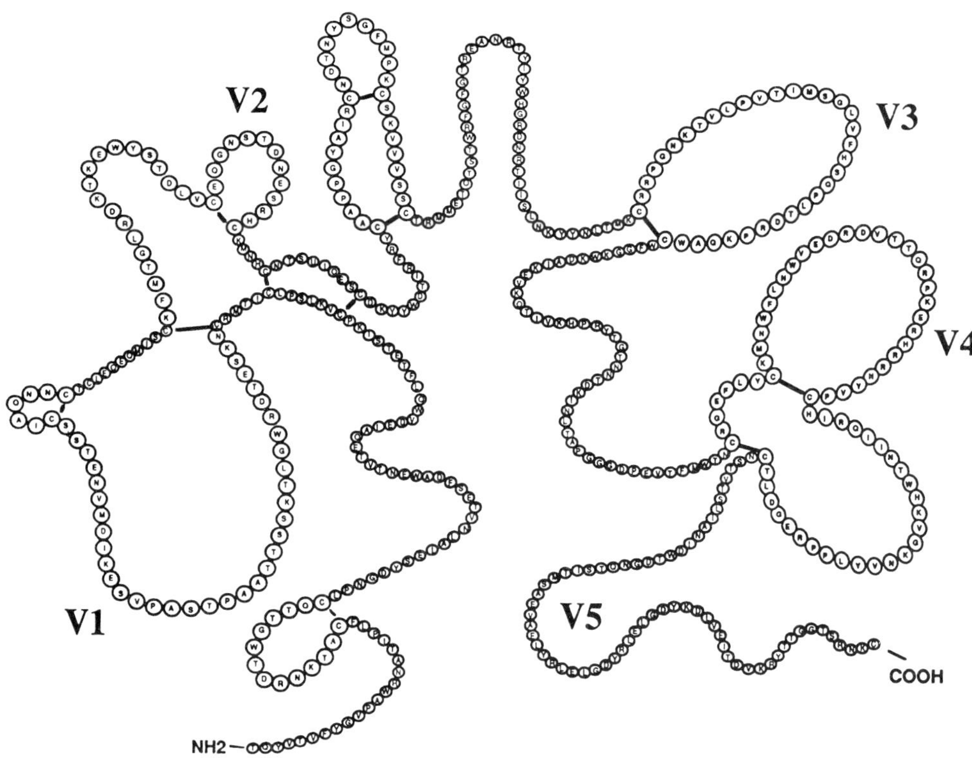

Fig. 1. SIV outer envelope protein (gp110). The diagram is based on disulfide analysis of HIV gp120 [6].

envelope. It was expressed using the baculovirus/insect cell system and is secreted into the medium [22]. Purification was achieved by a single affinity step utilizing infected macaque sera. On the basis of disulfide analysis of HIV gp120 (external envelope), a structure for gp110 was proposed compatible with the disulfide findings [6, 23, 24]. Since cysteines are also conserved between the HIV and SIV envelopes, one can draw an analogous diagram for the SIV gp110 (Mac251 isolate) assuming the same disulfide pattern. The results are shown in figure 1. With analogy to gp120, hypervariable domains V1–V5 can be identified in the SIV envelope. However, there are a number of important differences. One is that the envelope protein in SIV in general demonstrates less variability in sequences than those of HIV-1. The other difference is that the region in the SIV envelope that is analogous to the HIV V3 domain does not show much variability. Instead, the V4 region in SIV appears to be more highly variable [25]. In order to illustrate this point, we have presented some of the known sequences of SIV in the region of V3 and V4 for different SIV isolates (fig. 2). The envelopes of two SIV viral isolates, Mac251 and delta B670, have been studied in more detail in our laboratory. The sequence of the B670 isolate has been determined and will be published soon [G.J. LaRosa, in preparation]. Aside from lack of high variability in the V3 of SIV, the other significant difference between the V3 of the isolates occurs in the change from the positively charged amino acid lysine (K) to negatively charged aspartic and glutamic (D and E). Clearly, the more dramatic sequence variations are observed in the V4 region, pointing to the strong possibility that this domain is involved in defining the neutralization epitopes. This change in domain variability is probably related to the nature of the sequences of the V3 loops in the HIV and SIV envelope proteins. The HIV V3 is hydrophilic and is likely exposed, whereas in SIV the V3 is hydrophobic and presumably interacts with other domains in the envelope.

SIV gp110 binds to recombinant CD4 in a similar fashion to HIV gp120 (fig. 3). The binding is not inhibited by monoclonal OKT4, whereas the monoclonal OKT4A inhibits the complex formation between the envelope and CD4 [22]. We have also investigated the binding constant of gp110 to the recombinant CD4, using a Scatchard analysis (data not shown). In order to achieve this, we used a series of concentrations of CD4 and gp110 proteins. Using monoclonal OKT4, the complex of gp110-CD4 was precipitated on protein A-Sepharose. By employing the Western blotting technique, a mixture of polyclonal sera to CD4 and gp110 was used as primary antibodies followed by [^{125}I]-protein A treatment. The individual bands were visualized for both the immunoprecipitates and the supernatants. By scanning the film, one can calculate the concentration of the bound and unbound gp110. A value of 1.8×10^{-7} was obtained for K_d of SIV gp110 to CD4, which agrees with the results published by other investigators [26]. The binding constants for HIV-2 gp120 and SIV gp110 to CD4 are approximately 25–50 times lower than that of HIV gp120 [26, 27].

We also investigated the effect of glycosylation of SIV gp110 upon binding to CD4. The results are shown in figure 4. Glycosylated gp110 binds CD4 and is immunoprecipitated by monoclonal OKT4, whereas the deglycosylated gp110 does not form such a complex. This is in contrast to deglycosylated gp120 which still retains binding to CD4 (data not shown), in agreement with the results of another report [28]. Denaturation of SIV gp110 in the deglycosylation process appears to be an unlikely explanation for lack

```
                                                                                V3
                                                                        ______________________________
SmmB670-45k   265 TQTSTWFGFNGTRAENRTYIYWHGRSDRTIISLNKFYNLTIKCRRPGNKTVLPVTIMSGLVFHSQPINKRPKQAWCWFGGSWKKAIQEVKETIVKHPRYT
Smmh4-45k     279 ------------------------K-N--------Y----MR----E---------------------E---------E-------------L-------
Mm251-45k     273 -------------------------DN--------Y----M---------------------------D-----------K--D--K---Q---------
Mm239-45k     272 -------------------------DN--------Y----M---------------------------D-----------K--D--K---Q---------
Mm142-45k     273 -------R-----------------DN--------H----M-----------------A-------V-E-------R---N--E--K---Q---------
Mne-45k       272 -----------------------SKDN--------Y----M---------------------------D-------R-E-N--E--K---Q---------

                                                                       V4
                                                             ________________________________
SmmB670-45k   365 GTNDTKKINLTAPRGGDPEVTFMWTNCRGEFLYCKMNWFLNWVEDRDTNGSIWKEQKRKEQEKRNYVPCHIRQIVNTWHKVGRNVYLPPREGDLTCNSTV
Smmh4-45k     379 -----R-------A---------------------------------QK-GR--Q-N----Q-K----------I-------K-----------------
Mm251-45k     373 ---N-D-------G---------------------------------V..T...T-RP--RHR-----------I-------K-----------------
Mm239-45k     372 ---N-D-------G--------------------------------N-..A...N--P---H------------I-------K-----------------
Mm142-45k     373 ---NSD----------------------------------------SL..T...T--P--RH------------I-------K-----------------
Mne-45k       372 ---N-D-------G-------------------------------KNLT-T...T--PQ-RH------------I-------------------------

SmmB670-45k   465 TSLIANIDWIDGNQTNITMSAEVAELYRLELGDYKLVEITPIGLAPTNVRRYTTTGASRNKR
Smmh4-45k     479 -----E----NS-E----------------------I----------S--------------
Mm251-45k     468 ---------T-----S-------------------------------D-K----G-T-----
Mm239-45k     467 -----------------------------------------------D-K----G-T-----
Mm142-45k     468 -------N-T-----S---------------------------------K----G-T-----
Mne-45k       469 -------------------------------------------------K----G-T-----
```

Fig. 2. Sequence comparison of SIV gp110 from different virus isolates. The sequences correspond to the C-terminal portion of the envelope called 45-kD fragment (see text). The regions designated as V3 and V4 are identified.

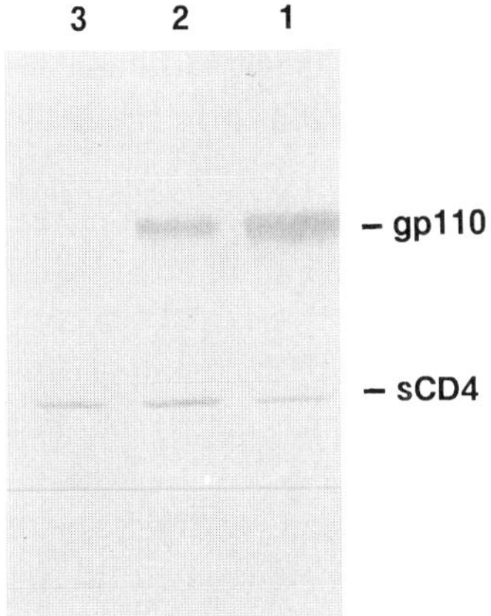

Fig. 3. Binding of recombinant SIV gp110 to soluble CD4 (sCD4). Immunoprecipitation was performed as described earlier [22]. Lane 1 = gp110 and sCD4 molecular markers; lane 2 = gp110 mixed with sCD4 in the presence of monoclonal antibody OKT4 followed by immunoprecipitation; lane 3 = same as lane 2, except monoclonal OKT4A was employed.

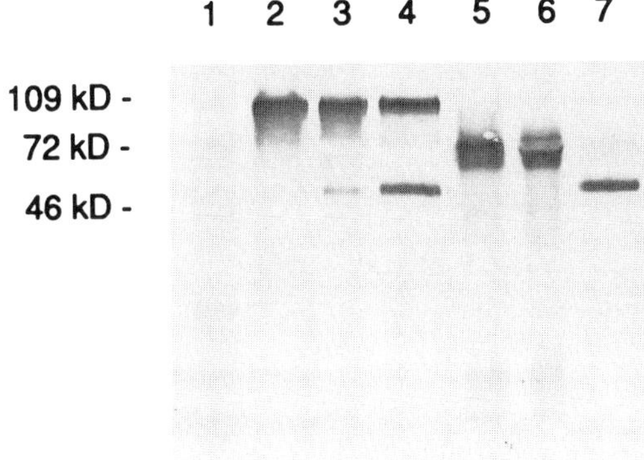

Fig. 4. Effect of deglycosylation of SIV gp110 on binding soluble CD4 (sCD4). Lane 1 = molecular weight markers; lane 2 = SIV gp110 marker; lanes 3 and 4 = immunoprecipitation of gp110 and sCD4 in the presence of monoclonal OKT4 (lane 3 is the supernatant and lane 4 is the precipitate); lane 5 = deglycosylated gp110 marker; lanes 6 and 7 = immunoprecipitation of deglycosylated gp110 and sCD4 in the presence of OKT4 (lane 6 is the supernatant and lane 7 is the precipitate). The protein band at 50 kD corresponds to sCD4 from baculovirus.

of binding of the SIV envelope to CD4. Recently, it has been reported that the loss of a single glycosylation site at amino acid 400 of the HIV-2 envelope was sufficient to reduce CD4 binding by at least 50 times [29].

Neutralization Domain of the SIV Envelope

In order to locate the neutralizing region of the SIV envelope, we adopted a similar strategy to that used with HIV-1 [1, 5]. We prepared envelope proteins and recombinant and synthetic fragments thereof from SIV-Mac251. Animals (goats, guinea pigs, rabbits) were immunized with these immunogens and their sera were analyzed for neutralizing titers. A map of the SIV envelope protein along with the recombinant subunits and synthetic peptides is presented in figure 5. Envelope protein gp140 contains the complete envelope except the C-terminal tail of the cytoplasmic domain. Both gp140 and gp110 (outer envelope protein) were expressed in the baculovirus/insect cell system. gp140 was purified under denaturing conditions from the cell pellet, whereas gp110 was purified from the medium using nondenaturing reagents. In addition, three other recombinant subunit pro-

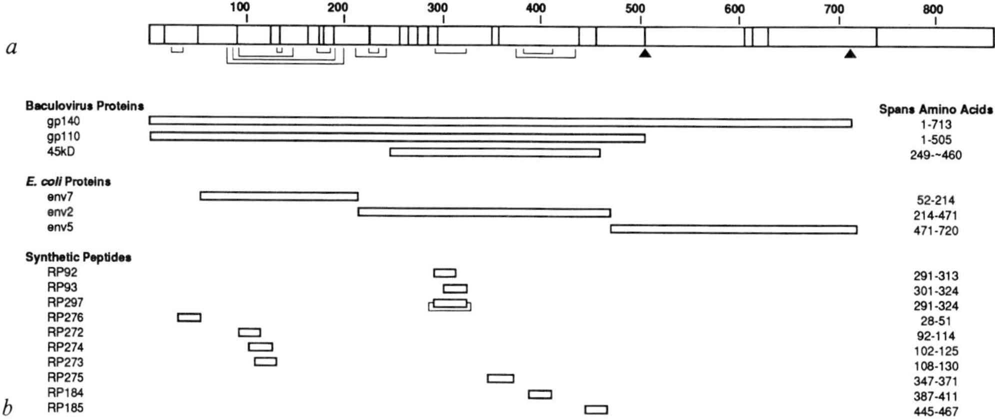

Fig. 5. Map of the SIV-Mac251 envelope protein *(a)* and location of recombinant envelope subunit proteins and synthetic peptides *(b)*.

teins, env2, env5 and env7, which together span the whole envelope, were expressed in *Escherichia coli* and purified using denaturing reagents. Finally, a series of peptides corresponding to regions of the envelope that might have been candidates for a neutralizing domain, based on our experience with HIV-1, were synthesized. In particular, three peptides related to the V3 loop of SIV (RP92, 93 and 297) analogous to the HIV V3 were prepared. One of them, RP297, has the complete amino acid sequence of the SIV V3 with a closed disulfide bridge. Sera obtained from animals immunized with these antigens were analyzed for ELISA and neutralization activities [22]. The data are shown in table 1. The data clearly demonstrate that only the native gp110 elicited neutralizing antibodies. None of the synthetic peptides or denatured proteins produced neutralizing sera. The low neutralizing titer observed with gp140 is probably due to some refolding of the envelope molecule. The results shown in table 1 lead us to conclude that the neutralizing epitope(s) of the SIV envelope are conformation-dependent and the tertiary structure of the native molecule is required to elicit neutralizing antibodies. These results are in contrast to HIV-1 where small linear peptides derived from the V3 domain of the HIV-1 envelope produced high neutralizing sera. As it was pointed out earlier, the PND of HIV-1 is conformation-independent and small linear peptides from this domain have been shown to completely block the neutralizing activity associated with

Table 1. ELISA and SIV-Mac251 neutralization titers of immune sera

Immunogen	Type of animal	ELISA titer	Titer in synctium reduction assay	Titer in cell viability assay
Mac251-infected	macaque	7,000	45,000	ND
		9,000	800,000	40,960
Denatured gp140	guinea pig	60,000	42	80
		≥ 109,000	≤20	160
		≥ 109,000	≤ 20	40
	goat	≥ 109,000	28	ND
		≥ 109,000	120	160
Native gp110	guinea pig	≥ 109,000	12,500	ND
		≥ 109,000	15,500	5,120
		≥ 109,000	37,000	5,120
	goat	≥ 109,000	46,000	ND
		≥ 109,000	80,000	ND
Denatured gp110	guinea pig	≥ 109,000	≤ 10	ND
		60,000	≤ 10	ND
		97,000	≤ 10	ND
env7	guinea pig	30,000	≤ 20	≤ 20
		25,000	≤ 20	≤ 20
		60,000	≤ 20	≤ 20
env2	guinea pig	70,000	≤ 20	≤ 20
		11,000	≤ 20	≤ 20
		35,000	≤ 20	≤ 20
env5	guinea pig	≥ 109,000	≤ 20	≤ 20
		≥ 109,000	≤ 20	≤ 20
		≥ 109,000	≤ 20	≤ 20
RP92	guinea pig	40,000	≤ 20	≤ 20
		20,000	≤ 20	≤ 20
		4,500	≤ 20	≤ 20
RP93	guinea pig	95,000	≤ 20	≤ 20
		95,000	≤ 20	≤ 20
		55,000	≤ 20	≤ 20
RP297, 184, 185,	guinea pig		≤ 20	ND
272, 273, 274			≤ 20	ND
			≤ 20	ND

gp160 antisera [1]. As a further test for investigating the conformational characteristics of the neutralizing domain of SIV, we carried out studies to test the inhibition of neutralization of anti-gp110 serum and infected macaque serum by different proteins, protein subunits and peptides [22]. The

Table 2. Inhibition of neutralization of anti-native gp110 serum and infected macaque serum by native gp110 and gp110-derived proteins

Addition	Concentration *M*	Percent inhibition of gp110 sera	Percent inhibition of macaque sera
Native gp110	7×10^{-8}	99.4	99.8
Denatured gp110	7×10^{-8}	0	62
Denatured gp140	5×10^{-8}	18	50
Deglycosylated gp110	7×10^{-8}	87	98.2
V8 digest of gp110	7×10^{-8}	96.4	99.5
45-kD fragment	20×10^{-8}	93.2	98.2
RP297	1.5×10^{-6}	0	0
RP297, 272, 273, 184, 185, 274, 275, 276	3×10^{-6}	0	0

results of such a competitive study are presented in table 2. Only native gp110 was capable of blocking the neutralizing activities. V8 digestion of native gp110 resulted in a mixture of peptides that still blocked the neutralizing sera. In particular, a 45-kD fragment from this V8 digestion, which was purified by HPLC, was found to be responsible for this blockade. N-terminal amino acid analysis of this 45-kD peptide led us to propose the diagram shown in figure 6 for this molecule. V3, V4 and CD4-binding domains are included in this fragment.

In another study a group of investigators used a mammalian cell culture system for expressing SIV-Mac envelope protein [30]. The glycosylated protein called rgp130 must be similar to the SIV gp110 described here except the glycosylation patterns may not be identical. Rhesus macaques were immunized with rgp130. The protein did not elicit neutralizing antibodies. We have carried out a similar immunization protocol with baculovirus gp110 and in all cases high neutralizing titers were observed (data not shown). The reason for this discrepancy is not clear, although it may be due to different neutralization assays employed.

Böttiger et al. [31] tested a total of 42 HIV-2-positive sera from humans for cross-reactivity between HIV-1 and HIV-2. The authors concluded that there was a correlation between neutralization of HIV-2 and reactivity to a 35-amino-acid peptide, representing the V3 loop of HIV-2. In view of the similarity between SIV and HIV-2, it remains to be found whether the neutralizing determinants in HIV-2 are conformation-dependent like those of SIV.

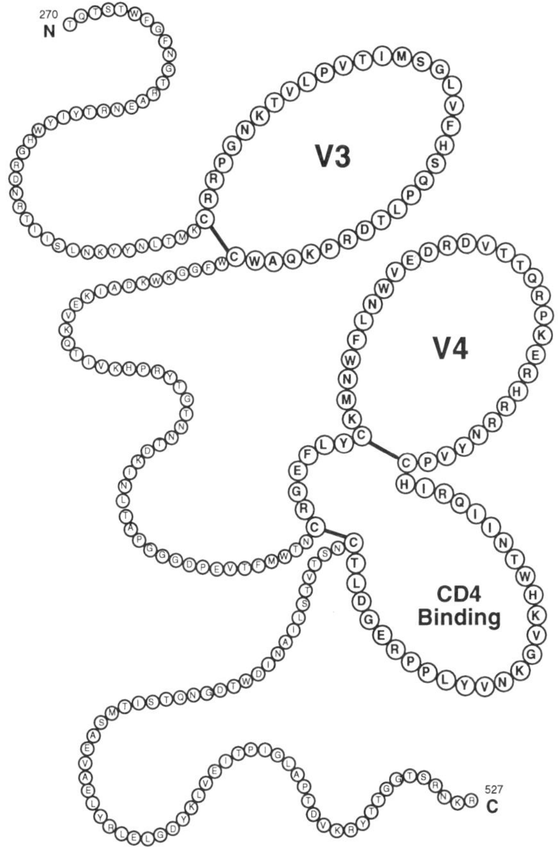

Fig. 6. Diagram representing the structure of a 45-kD fragment resulting from V8 protein digestion of SIV gp110. It was purified from HPLC. The N-terminal amino acid analysis was carried out to locate the N terminus.

A number of neutralizing monoclonals have been produced that map to the SIV envelope. Monoclonal SF8/5E11 recognizes the transmembrane portion of SIV-Mac251 and is isolate-specific [32]. The determinant corresponds to a variable, weak neutralizing epitope. Eighteen monoclonal antibodies to the SIV-Mac251 envelope were prepared by Kent et al. [33]. Some of them were found to be neutralizing. In particular, two of them, KK5 and KK9, are conformation-dependent and demonstrated high neutralizing titers. They are being investigated at the present time in our laboratory to learn more about the neutralizing epitopes of the SIV envelope.

Conclusion

In contrast to HIV-1, the PND of the SIV envelope is conformation-dependent. It appears that a number of the segments of the envelope, including disulfide loops, are involved in defining this domain. The transition from conformation-dependent (SIV) to conformation-independent (HIV) neutralizing determinants raises a number of questions that need to be explored. With regard to the suitability of an SIV vaccine as a model for HIV, certain caution must be exercised. Clearly the nature of the neutralization determinants is different between the two viruses. However, if neutralizing titers prove to be important in protection, one may still conclude that neutralization per se is the source of protection, not the type of determinants eliciting neutralizing antibodies. Undoubtedly, SIV will serve as an important model for learning about HIV.

References

1 Javaherian K, Langlois AJ, McDanal C, Ross KL, Eckler LI, Jellis CJ, Profy AT, Rusche JR, Bolognesi DP, Putney SD, Matthews TJ: Principal neutralizing domain of the human immunodeficiency virus type 1 envelope protein. Proc Natl Acad Sci USA 1989;86:6768–6772.

2 Goudsmit J, Debouck C, Meloen RH, Smit L, Bakker M, Asher DM, Volff AV, Gibbs CJ Jr, Gajdusek DC: Human immunodeficiency virus type 1 neutralization epitope with conserved architecture elicits early type-specific antibodies in experimentally infected chimpanzees. Proc Natl Acad Sci USA 1988;85:4478–4482.

3 Palker TJ, Clark ME, Langlois AJ, Matthews TJ, Weinhold KJ, Randall RR, Bolognesi DP, Haynes BF: Type-specific neutralization of the human immunodeficiency virus with antibodies to *env*-encoded synthetic peptides. Proc Natl Acad Sci USA 1988;85:1932–1936.

4 Emini EA, Schleif WA, Nunberg JH, Conley AJ, Eda Y, Tokiyoshi S, Putney SD, Matsushita S, Cobb KE, Jett CM, Eichberg JW, Murthy KK: Prevention of HIV-1 infection in chimpanzees by gp120 V3 domain-specific monoclonal antibody. Nature 1992;355:728–730.

5 Rusche JR, Javaherian K, McDanal C, Petro J, Lynn DL, Grimaila R, Langlois AJ, Gallo RC, Arthur LO, Fischinger PJ, Bolognesi DP, Putney SD, Matthews TJ: Antibodies that inhibit fusion of human immunodeficiency virus-infected cells bind a 24-amino acid sequence of the viral envelope, gp120. Proc Natl Acad Sci USA 1988;85:3198–3202.

6 Leonard CK, Spellman MW, Riddle L, Harris RJ, Thomas JN, Gregory TJ: Assignment of intrachain disulfide bonds and characterization of potential glycosylation sites of the type 1 recombinant human immunodeficiency virus envelope glycoprotein (gp120) expressed in Chinese hamster ovary cells. J Biol Chem 1990; 265:10373–10382.

7 Javaherian K, Langlois AJ, LaRosa GJ, Profy AT, Bolognesi DP, Herlihy WC, Putney SD, Matthews TJ: Broadly neutralizing antibodies elicited by the hypervariable neutralizing determinant of HIV-1. Science 1990;250:1590–1593.
8 LaRosa GJ, Davide JP, Weinhold K, Waterbury JA, Profy AT, Lewis JA, Langlois AJ, Dreesman GR, Boswell RN, Shadduck P, Holley LH, Karplus M, Bolognesi DP, Matthews TJ, Emini EA, Putney SD: Conserved sequence and structural elements in the HIV-1 principal neutralizing determinant. Science 1990;249:932–935.
9 Takahashi H, Nakagawa Y, Pendleton CD, Houghten RA, Yokomura K, Germain RN, Berzofsky JA: Induction of broadly cross-reactive cytotoxic T cells recognizing an HIV-1 envelope determinant. Science 1992;255:333–336.
10 Weiss RA, Clapham PR, McClure MO, McKeating JA, McKnight A, Dalgleish AG, Sattentau QJ, Weber JN: Human immunodeficiency viruses: Neutralization and receptors. J AIDS 1988;1:536–541.
11 Steimer KS, Scandella CJ, Skiles PV, Haigwood NL: Neutralization of divergent HIV-1 isolates by conformation-dependent human antibodies to gp120. Science 1991;254:105–108.
12 Hu S-L, Klaniecki J, Dykers T, Sridhar P, Travis BM: Neutralizing antibodies against HIV-1 BRU and SF2 isolates generated in mice immunized with recombinant vaccinia virus expressing HIV-1 (BRU) envelope glycoproteins and boosted with homologous gp160. AIDS Res Hum Retroviruses 1991;7:615–620.
13 Ho DD, McKeating JA, Li XL, Moudgil T, Daar ES, Sun N-C, Robinson JE: Conformational epitope on gp120 important in CD4 binding and human immunodeficiency virus type 1 neutralization identified by a human monoclonal antibody. J Virol 1991;65:489.
14 Posner MR, Hideshima T, Cannon T, Mukherjee M, Mayer KH, Byrn RA: An IgG human monoclonal antibody that reacts with HIV-1/gp120, inhibits virus binding to cells, and neutralizes infection. J Immunol 1991;146:4325–4332.
15 Haigwood NL, Shuster JR, Moore GK, Lee H, Skiles PV, Higgins KW, Barr PJ, George-Nascimento C, Steimer KS: Importance of hypervariable regions of HIV-1 gp120 in the generation of virus neutralizing antibodies. AIDS Res Hum Retroviruses 1990;6:855–869.
16 Thali M, Olshevsky U, Furman C, Gabuzda D, Posner M, Sodroski J: Characterization of a discontinuous human immunodeficiency virus type 1 gp120 epitope recognized by a broadly reactive neutralizing human monoclonal antibody. J Virol 1991; 65:6188–6193.
17 Berkower I, Murphy D, Smith CC, Smith GE: A predominant group-specific neutralizing epitope of human immunodeficiency virus type 1 maps to residues 342 to 511 of the envelope glycoprotein gp120. J Virol 1991;65:5983–5990.
18 Haigwood NL, Nara PL, Brooks E, Van Nest GA, Ott G, Higgins KW, Dunlop N, Scandella CJ, Eichberg JW, Steimer KS: Native but not denatured recombinant human immunodeficiency virus type 1 gp120 generates broad-spectrum neutralizing antibodies in baboons. J Virol 1992;62:172–182.
19 Berman PW, Gregory TJ, Riddle L, Nakamura GR, Champe MA, Porter JP, Wurm FM, Hershberg RD, Cobbs EK, Eichberg JW: Protection of chimpanzees from infection by HIV-1 after vaccination with recombinant glycoprotein gp120 but not gp160. Nature 1990;345:622–628.
20 Gardner MB: Vaccination against SIV infection and disease. AIDS Res Hum Retroviruses 1990;6:835–845.

21 Hu S-L, Abrams K, Barber GN, Moran P, Zarling JM, Langlois AJ, Kuller L, Morton WR, Benveniste RE: Protection of macaques against SIV infection by subunit vaccines of SIV envelope glycoprotein gp160. Science 1992;255:456–459.
22 Javaherian K, Langlois AJ, Schmidt S, Kaufmann M, Cates N, Langedijk JPM, Meloen RH, Desrosiers RC, Burns DPW, Bolognesi DP, LaRosa GJ, Putney SD: The principal neutralization determinant of simian immunodeficiency virus differs from that of human immunodeficiency virus type 1. Proc Natl Acad Sci USA 1992;89: 1418–1422.
23 Gregory T, Hoxie JA, Watanabe C, Spellman M: Structure and function in recombinant HIV-1 gp120 and speculation about the disulfide bonding in gp120 homologs 1 HIV-2 and SIV; in Atassi M (ed): Advances in Experimental Medicine and Biology. Proc Int Symp on Immunobiology of Proteins and Peptides. New York, Plenum, 1991, pp 1–14.
24 Hoxie JA: Hypothetical assignment of intrachain disulfide bonds for HIV-2 and SIV envelope glycoprotein. AIDS Res Hum Retroviruses 1991;7:495–499.
25 Overbaugh J, Rudensey LM, Papenhausen MD, Benveniste RE, Morton WR: Variation in simian immunodeficiency virus *env* is confined to V1 and V4 during progression to simian AIDS. J Virol 1991;65:7025–7031.
26 Ivey-Hoyle M, Culp JS, Chaikin MA, Hellmig BD, Matthews TJ, Sweet RW, Rosenberg M: Envelope glycoproteins from biologically diverse isolates of immunodeficiency viruses have widely different affinities for CD4. Proc Natl Acad Sci USA 1991;88:512–516.
27 Moore JP: Simple methods for monitoring HIV-1 and HIV-2 gp120 binding to soluble CD4 by enzyme-linked immunosorbent assay: HIV-2 has a 25-fold lower affinity than HIV-1 for soluble CD4. AIDS 1990;4:297–305.
28 Fenouillet E, Clerget-Raslain B, Gluckman JC, Guétard D, Montagnier L, Bahraoui E: Role of *N*-linked glycans in the interaction between the envelope glycoprotein of human immunodeficiency virus and its CD4 cellular receptor. J Exp Med 1989;169: 807–822.
29 Morikawa Y, Moore JP, Wilkinson AJ, Jones IM: Reduction in CD4 binding affinity associated with removal of a single glycosylation site in the external glycoprotein of HIV-2. Virology 1991;180:853–856.
30 Planelles V, Haigwood NL, Marthas ML, Mann KA, Scandella C, Lidster WD, Shuster JR, Van Kuyk R, Marx PA, Gardner MB, Luciw PA: Functional and immunological characterization of SIV envelope glycoprotein produced in genetically engineered mammalian cells. AIDS Res Hum Retroviruses 1991;7:889–897.
31 Böttinger B, Karlsson A, Andreasson P-A, Nauclér A, Costa CM, Norrby E, Biberfeld G: Envelope cross-reactivity between human immunodeficiency virus types 1 and 2 detected by different serological methods: Correlation between cross-neutralization and reactivity against the main neutralizing site. J Virol 1990;64:3492–3499.
32 Kodama T, Burns DPW, Silva DP, DiMarzo Veronese F, Desrosiers RC: Strain-specific neutralizing determinant in the transmembrane protein of simian immunodeficiency virus. J Virol 1991;65:2010–2017.
33 Kent KA, Gritz L, Stallard G, Cranage MP, Colligan C, Thiriart C, Corcoran T, Silvera P, Stott EJ: Production and characterization of monoclonal antibodies to simian immunodeficiency virus envelope glycoproteins. AIDS 1991;5:829–836.

Kashi Javaherian, Repligen Corporation, One Kendall Square,
Cambridge, MA 02139 (USA)

Norrby E (ed): Immunochemistry of AIDS.
Chem Immunol. Basel, Karger, 1993, vol 56, pp 91–111

Epitopes of HIV-1 Glycoproteins Recognized by the Human Immune System

Suman Laal[a], *Susan Zolla-Pazner*[a, b]

[a] Research Center for AIDS and HIV Infection, New York Veterans Affairs Medical Center, and [b] The Department of Pathology and Center for AIDS Research, New York University Medical Center, New York, N.Y., USA

Early studies with sera from patients with acquired immunodeficiency syndrome and individuals seropositive for human immunodeficiency virus (HIV) suggested that the human immune system reacted strongly to the envelope glycoproteins, gp120 and gp41, of HIV. While humoral immune responses develop to other proteins of the virus as well, antibodies to gp120 and gp41 glycoproteins were primarily shown to mediate protective functions such as inhibition of virus-host cell fusion [1], neutralization of infectivity of cell-free virus [2–7] and induction of antibody-dependent cell-mediated cytotoxicity (ADCC) against infected cells [8, 9]. These antigens are also involved in generating cell-mediated immune responses to HIV, and thus play a significant role in determining the course of the infection [10]. However, in addition to their role in these protective responses, gp120 and gp41 have been known to participate in responses detrimental to the host, for example, by triggering autoimmune responses [11], inhibiting natural killer cell activity [12] and enhancing HIV infectivity [13–17]. This diversity of responses generated by gp120 and gp41 has led to attempts to characterize the relevant epitopes on these molecules and delineate their roles in the immune response.

Epitopes on gp120

Modrow et al. [18] compared the predicted amino acid sequences of HIV-1 envelope proteins from seven different isolates based on nucleotide sequences. They found that, although the amino acid sequences differed

Table 1. Epitopes on gp120

Computer-predicted epitopes	Epitopes identified by serological analysis using synthetic peptides		
Modrow et al. [18]	Goudsmit [19]	Palker et al. [20]	Neurath et al. [21]
	55–65		61–90
137–154	135–148		138–164
			164–187
186–203			
232–246			219–245
300–320	307–321	303–321	306–338
358–375			
394–412			
445–458			
459–469			
470–483			477–508
		504–518	

Amino acid position of epitopes according to the HXB2 sequence in Myers et al. [22].

substantially between different isolates, the envelope protein could be broadly subdivided into five variable regions (V1–V5) with 25% or fewer conserved residues and five constant regions (C1–C5) with greater than 75% conservation. They used a computer program to predict the secondary protein structure and potential antigenic epitopes by superimposing values for hydrophilicity, surface probability and flexibility. This method identified nine candidate epitopes in the gp120 molecule. These epitopes were localized to amino acids 137–154, 186–203, 232–246, 300–320, 358–375, 394–412, 445–458, 459–469 and 470–483. These epitopes were generally located in predicted β-turn regions of the protein and consisted of amino acid sequences that tend to be nonhydrophobic or flexible and hence have a higher probability of being exposed at the surface of the gp120 molecule. Epitopes 232–246, 358–375, 445–458 and 470–483 are located in the constant regions of gp120; the others are in the variable regions.

Mapping of epitopes that are recognized as part of the human immune response was described by Goudsmit [19] who used 536 overlapping nonapeptides derived from the gp120 nucleotide sequence of the HIV-IIIB BH10 clone and sera from infected individuals. This author identified three distinct immunodominant regions in the gp120 molecule (table 1). The first region spanned amino acids 55–65 in C1, the second, amino acids 135–148 in V1 and the third, amino acids 307–321 in V3 [22]. Work with these peptides failed to reveal another immunodominant area spanning amino acids 504–518 in the C4/C5 region identified by using longer peptides (15 amino acids) by Palker et al. [20]. Thus, of the nine computer-predicted immunodominant regions, only two were confirmed serologically, both in variable regions (V1 and V3) of gp120. However, the sera identified two immunodominant areas in the constant (C1 and C4/5) regions of gp120, not predicted by the computer model.

Further peptide-based, human serological analysis of epitopes of gp120 was performed by Neurath et al. [21] who used longer peptides (19–36 amino acids) on the assumption that these were more likely to mimic the three-dimensional conformations present in native gp120. Eighteen peptides spanning the entire $gp120_{IIIB}$ molecule were studied, and although 16 of these were found to be immunogenic in rabbits, human HIV-1-positive sera reacted with only six. These included peptides 61–90, 138–164, 164–187, 219–245, 306–338 and 477–508. Of these, four epitopes overlapped with those predicted by computer analysis and/or with those identified using shorter peptides, and two were previously unrecognized (table 1).

The involvement of the V3 region of gp120 in eliciting neutralizing antibodies has been suggested by several studies [7, 23–25]. This region starts at amino acids 296–309 and ends at amino acids 331–343, depending on the HIV-1 strain, and forms a loop due to the disulfide bonding between the cysteines at each end. Amino acid analysis of this epitope in a large number of isolates from North America and Europe has shown conserved features of this loop [26]. The amino acids adjacent to the cysteines within the loop are quite conserved as is the central tetrapeptide at the tip of the loop. This tetrapeptide displays limited variation in isolates from around the world.

Recognition of V3 epitopes during infection has been characterized by Gorny et al. [2], Scott et al. [27] and Tilley et al. [28], using human monoclonal antibodies (MoAbs) derived from infected individuals. In the experiments by Gorny et al. [2], peripheral blood mononuclear cells (PBMCs) were first transformed with Epstein-Barr virus (EBV) and later fused with a mouse × human heteromyeloma (SHM-D33 [29]). Hybrids were cloned repeatedly at 100–1 cell per well until stable, monoclonal hydridomas were obtained. Screening at every stage was performed with a 23-amino acid peptide that spans the gp120 V3 loop of the MN strain of HIV (YNKRKRI-HIGPGRAFYTTKNIIG). This site-selective approach yielded several clones making antibodies, some of which have been characterized for their

Table 2. Human MoAbs to epitopes in the V3 region

MoAb	Epitope on $V3_{MN}$	Cross-reactivity	Affinity for V3 (K_D $10^{-6}M$)		Neutralization (50%) (μg/ml)
			MN	SF2	
447-D	GPGR	MN, SF2, NY5, CDC4, RF, WMJ2, HXB2	0.56	0.32	MN (0.052) IIIB (1.8)
537-D	IGPGR	MN, SF2, NY5, CDC4, RF, WMJ2	1.3	8.5	MN (2.7)
268-D	HIGPGR	MN, SF2, NY5, CDC4, RF	0.59	2.3	MN (0.1)
386-D	HIGPGR	MN, SF2, NY5, CDC4, RF	0.18	0.85	MN (0.117)
477-D	HIGP	MN, SF2	ND	ND	
419-D	IHIGPGR	MN, SF2, NY5, WMJ2	3.8	0.23	MN (0.99)
453-D	IHIGPGR	MN, SF2, NY5, RF	1.6	3.1	MN (0.109)
504-D	IHIGPGR	MN, SF2	0.35	0.83	MN (0.728)
311-D	KRIHIGP	MN, SF2	7.4	6.0	MN (1.97)
257-D	KRIHI	MN, SF2, NY5, RF	0.23	0.22	MN (0.012)
694/98D	GRAF	MN, SF2, NY5, CDC4, RF, WMJ2, HXB2	0.1	1.8	MN (0.46) IIIB (0.04)
418-D	HIGPGRA	MN	0.24	–	MN (0.014)
391/95D	RKRIHIGPGRAFYTT	MN, SF2	0.91	4.4	MN (0.04)
412-D	RKRIHIGPGRAFYTT	MN	9.1	–	MN (4.69)
N701.9b	PGRAFYT[1]	MN, SF2, J62	–	–	MN (0.90)[1]
4117C	IXIGPGR[1]	MN, SF2, RF, JR-CSF, 3/15 African isolates	0.9×10^9L/M		MN (0.7)[1]

[1] Methods used are not necessarily the same as those used for all the other MoAbs [2, 30].

immunochemical and biological activity [30]. One MoAb directed against the V3 region of MN, produced by Scott et al. [27] and Robinson et al. [31], was screened against concanavalin-A-immobilized HIV glycoproteins from detergent-disrupted supernatant fluids from HIV-1-producing cell lines. Another MoAb has been described by Tilley et al. [28], who used recombinant $gp160_{BRU}$ for the initial screen.

All 14 MoAbs obtained in our laboratory [2, 30] reacted with $V3_{MN}$ (YNKRKRIHIGPGRAFYTTKNII) and twelve of these also reacted with the closely related $V3_{SF2}$ peptide (NNTRKSIYIGPGRAFHTTGRII); two MoAbs (418-D and 412-D) were specific for MN (table 2). All anti-V3

MoAbs reacted with both native and reduced (dithiothreitol-treated and alkylated) MN lysate, indicating recognition of linear epitopes. Epitope mapping of these MoAbs was done using overlapping hexa- or heptapeptides. Twelve of the 14 MoAbs were found to recognize epitopes within a limited region (KRIHIGPGRAF) of the $V3_{MN}$ loop (table 2). Two MoAbs, 391/95D and 412-D, showed reactivity only with 15-mer peptides and not with the smaller peptides tested.

The most broadly cross-reactive MoAb was 447-D, which reacted with V3 peptides of seven of the eight HIV strains tested, and neutralized strains as divergent as HIV-MN and HIV-IIIB. Epitope mapping of this antibody with overlapping hexapeptides showed that 447-D recognized the peptides HIGPGR, IGPGRA and GPGRAF, suggesting that GPGR represents the core of the epitope to which this MoAb binds. All seven V3 peptides that 447-D reacted with possessed the GPGR sequence, while the remaining V3 peptide (ELI), to which 447-D failed to bind, has the sequence GLGQ instead of GPGR [32]. Further analysis was performed using hexamers of the parent peptide, HIGPGR, substituted with each amino acid at each position; these studies showed that H, I and the second G could be substituted by several different amino acids, whereas the G, P and R were critical for reactivity; thus, the core epitope is more accurately described as GP$\underline{X}$R. However, analysis of dissociation constants (K_D) for interaction of MoAb 447-D with the V3 peptides from four different strains [30] revealed that, in addition to the central GPGR motif, the flanking amino acids do contribute significantly to the binding of antibody 447-D. Thus, the binding affinity of 447-D is significantly lower for the V3 peptide of IIIB (HXB2) than for MN or SF2. It is noteworthy that the results obtained from peptide studies correlate well with neutralization studies in that the MN isolate requires significantly less antibody than IIIB for 50% neutralization in a syncytium inhibition assay.

Most of the epitopes recognized by the human anti-$V3_{MN}$ MoAbs are situated to the left of the GPGR tip and only one antibody (257-D) recognizes an epitope (KRIHI) that is independent of the GPGR motif. The affinities of 257-D and 447-D for MN and SF2 V3 peptides are comparable, and they have similar neutralizing activity for MN. 257-D cross-reacts in ELISA with SF2, NY5 and RF strains, even though the $V3_{RF}$ peptide shares only one amino acid and the $V3_{SF2}$, three amino acids, with the KRIHI sequence. However, unlike 447-D, 257-D neither binds to the HXB2 peptide nor neutralizes HIV-IIIB.

Two MoAbs, 268-D and 386-D, recognize the same hexapeptide, HIGPGR, in an ELISA format, and both show a similar range of cross-reactivity. The affinity of these MoAbs for MN, SF2 and RF strains of HIV are also similar; however, the affinity of 268-D for $V3_{NY5}$ is significantly lower than that of 386-D.

Another set of MoAbs, 453-D, 419-D and 504-D, all recognize the heptapeptide IHIGPGR. However, 504-D fails to cross-react with the $V3_{NY5}$ peptide, whereas the other two MoAbs show binding to it. The affinity of 504-D for $V3_{MN}$ is 10-fold higher than that of 419-D and 453-D. Replacement analysis of individual amino acids in the HIGPGR and IHIGPGR 'parent' peptides may reveal different contact residues within the combining sites of these MoAbs and may account for the differences in reactivities observed in the affinity assays.

Only two MoAbs are directed towards epitopes that include amino acids towards the right of the GPGR tip. MoAb 694/98D recognizes the motif GRAF and, like 447-D, shows cross-reactivity with MN and HXB2 in ELISA, as well as cross-neutralization of MN and IIIB. MoAb 418-D recognizes the heptapeptide HIGPGRA, and shows no cross-reactivity, even with $V3_{SF2}$, despite the fact that the two differ in only one amino acid in the epitope sequence (H/Y).

Of the two MoAbs which show reactivity with only a 15-mer peptide (RKRIHIGPGRAFYTT), MoAbs 391/95D and 412-D, the former neutralizes MN at a very low concentration (40 ng/ml) whereas the latter requires 1,000-fold more antibody.

A human MoAb, N701.9b, described by Scott et al. [27], reacts specifically with gp120 from MN, SF2 and J62 strains (table 2). Epitope mapping using a series of 14-amino-acid peptides derived from the $V3_{MN}$ loop sequence indicated that this MoAb recognizes the loop sequence PGRAFYTT. Neutralization assays done with seven isolates of HIV showed that the isolates containing the PGRAFYTT sequence were neutralized at less than 1 μg/ml, while one isolate with a single amino acid difference (A/V) required 1 μg/ml. Viruses that had a GPGR motif, but differed in the rest of the epitope, failed to be neutralized.

The anti-V3 human MoAb, 4117C, described by Tilley et al. [28], recognizes a 7-amino-acid epitope IXIGPGR near the tip of the V3 loop (table 2). It reacts with MN, SF2, RF, JR-CSF and 3/15 African isolates tested, and neutralizes both MN and SF2. It fails to react with or neutralize HIV-IIIB.

Serological analysis of epitopes recognized by polyclonal neutralizing antibodies of infected individuals [33] had shown that, early after infection, isolate-restricted neutralizing antibodies appeared. These antibodies were later found chiefly to recognize linear determinants in the V3 region of gp120. Broadly neutralizing antibodies appear later during the course of infection, and are believed to be directed primarily against non-V3 epitope(s) of gp120 [34]. A primary candidate for such epitope(s) has been the CD4-binding region of gp120, especially since HIV-positive sera usually inhibit CD4-gp120 binding. Results from several groups suggest that the second, third and fourth conserved regions of gp120 participate in forming the CD4-binding domain [35, 36]. Thus, unlike the V3 region, where the epitopes are primarily linear, the CD4-binding region appears to be a large, conformational and discontinuous epitope [34]. Despite its complexity, several groups have reported antibodies that are directed against the CD4-binding region [34, 37–39].

A human MoAb that has the capacity to block gp120-CD4 interaction has been reported by Ho et al. [34]. This MoAb (15e) was obtained by screening supernatants of EBV-transformed cells from infected individuals against concanavalin-A-bound gp120 from disrupted viral particles. The MoAb 15e reacts with gp120 from IIIB, MN and Z84, but not from AL and RF strains of HIV (table 3). It neutralizes IIIB, Z84, MN and SA-3, but not AL and RF. Neutralization probably occurs by blocking the CD4-gp120 interaction. MoAb 15e does not react with nonglycosylated gp120 made by *Escherichia coli* or *Saccharomyces cerevisiae*, suggesting that carbohydrate residues are essential directly or indirectly for recognition. This MoAb does not bind to the V3 or C2 peptides, or to reduced gp120. Competitive binding studies with several murine MoAbs directed against the CD4-binding domain suggest that 15e recognizes an epitope that is not recognized by the murine MoAbs. Antibodies to the epitope recognized by 15e are induced several months after seroconversion in infected individuals.

Our laboratory has produced human MoAbs directed against the CD4-binding domain by screening EBV-transformed PBMCs for their ability to produce antibodies against $rgp120_{IIIB}$. Three MoAbs have been obtained, 559/64D, 588-D and 448-D [39]. Each recognizes conformational epitopes on gp120, since reduction abrogates the ability of rgp120 to bind these MoAbs in both ELISA and radioimmunoprecipitation assays. All three MoAbs cross-react with gp120 of MN, SF2 and IIIB strains. They fail to react with V3 peptides, but inhibit the interaction of rCD4 with rgp120 in a competitive binding assay. These data suggest that these MoAbs recognize epitope(s) in or near the CD4-binding region.

Table 3. Epitopes in non-V3 regions of gp120

Antibody	Epitope/region	Cross-reactivity	Neutralization	Affinity (K_D $10^{-8}M$)	Miscellaneous	Reference
15e	Conformational CD4-binding region	IIIB, MN, Z84	IIIB, Z84, MN, SA-3, 10 primary isolates	–	No binding with V3 peptides; carbohydrate-dependent epitope	Robinson et al. [31] Ho et al. [34]
559/64D	Conformational CD4-binding region	MN, SF2, IIIB	MN, SF2, IIIB, AL-1	4.0 (IIIB) 1.4 (SF2)	No binding with V3 peptides	Karwowska et al. [39] Gorny et al. [41]
588-D	Conformational CD4-binding region	MN, SF2, IIIB	MN, IIIB, RF, AL-1, SF-2	2.0 (IIIB) 0.6 (SF2)	No binding with V3 peptide	Karwowska et al. [39] Gorny et al. [41]
448-D	Conformational CD4-binding region	MN, SF2, IIIB	MN, SF2, IIIB, AL-1	2.9 (IIIB) 0.5 (SF2)	No binding with V3 peptide	Karwowska et al. [39] Gorny et al. [41]
S1-1	Conformational CD4-binding region	IIIB, MN, RF	IIIB, MN, RF	–		Lake et al. [40]
F105	Conformational CD4-binding region or nearby	IIIB, RF, MN,SF2	IIIB	–	–	Posner et al. [38]
1125H	Conformational CD4-binding region	IIIB, RF, MN, SF2, several early and late passages of primary isolates	IIIB, RF, MN, SF2	1.3×10^{9} L/M (BRN)	V3 cleavage does not interfere with epitope	Tilley[1] et al. [37]
450-D	Linear (487–508)	MN, SF2	no			Karwowska et al. [39]

[1] Methods used are not the same as those used for other MoAbs [39].

The ability of these antibodies to neutralize seven different strains of HIV has been assessed [41]. MoAb 448-D neutralizes SF2 and AL-1 at less than 0.1 μg/ml, and IIIB, at 6.25 μg/ml. 559/64D shows a similar pattern in that SF2, AL-1 and MN are neutralized at low concentrations, but WMJ-2, DU-7887-7 and RF are not neutralized even at concentrations above 50 μg/ml. In contrast, 588-D fails to neutralize MN below 6.25 μg/ml and, unlike 559-D, neutralizes RF. This suggests that, although these antibodies are directed against a conformational region grossly designated as the CD4-binding domain, in reality there are differences within the region that the antibodies perceive.

Some insight is provided by the studies involving identification of amino acids that are important for recognition of gp120 by these MoAbs [42]. This has been done by checking their reactivity with a set of HIV-1 gp120 mutants that have alterations in their conserved residues [36]. Amino acid changes at various positions have differential effects on the binding of MoAbs 448-D, 559-D and 588-D to the mutant gp120s when compared to binding to the wild-type gp120 [42]. The binding of MoAb 588-D was abolished when the mutants contained alterations at amino acids 256 and 262. Binding of MoAb 559-D was abolished when alterations occurred at amino acids 117, 256 and 262. Binding of MoAb 448-D was abolished by alterations of amino acids 113, 256 and 368, and was significantly reduced for mutants with changes at amino acids 88, 117, 257 and 262.

Another human MoAb with properties similar to the anti-CD4-binding domain antibodies described above has been reported by Tilley et al. [37]. This MoAb, 1125H, reacts with rgp120 and rgp160, but not with gp41 peptides. It shows broad cross-reactivity, reacting with cells infected with IIIB, RF, MN and SF2, as well as with early and late passages of several primary isolates of HIV. However, it fails to recognize HIV-2 (table 3) or several African isolates of HIV-1. As seen with other MoAbs directed against conserved epitopes, 1125H neutralizes divergent strains of HIV-1 (MN, RF, SF2, IIIB), although unlike 15e, it neutralizes RF. The conformational nature of the epitope is evident from the loss of reactivity of this MoAb with reduced gp160 and gp120. The epitope recognized by 1125H appears to be independent of the presence of carbohydrate residues in that removal of N-linked sugars from gp120 does not affect reactivity. The binding of 1125H to gp160 is inhibited by CD4 suggesting that the epitope is located in or near the CD4-binding region of gp120.

Posner et al. [38] have obtained a MoAb (F105) which was derived from a patient with advanced HIV-1 infection. F105 reacts with surface antigens expressed on H9 cells infected with IIIB, SF2, RF and MN, and it neutralizes HIV-IIIB infection apparently by inhibiting viral binding to HT-H9 cells (table 3). The epitope recognized by F105 appears to lie within or near the CD4-binding site since soluble rCD4 inhibits the binding of F105 to gp120 expressed on the surface of infected cells. F105 does not react with denatured gp120, indicating that it recognizes a conformational epitope.

Recently, Lake et al. [40] have described a human MoAb, S1-1, produced by fusing B cells derived from the spleen of a seropositive individual to a mouse myeloma cell line. The epitope recognized by S1-1 is conformational and is conserved among HIV-IIIB, -MN and -RF. It is expressed on the surface of infected cells. The antibody S1-1 is able to neutralize IIIB and MN independently of the presence of complement, whereas neutralization of RF and of a primary isolate requires complement. S1-1 inhibits the binding of gp120 to soluble rCD4, suggesting that the epitope it recognizes lies in or near the CD4-binding region.

Serological studies had also identified an immunodominant region in the C-terminal region of gp120 [21, 43]. However, only one human MoAb (450-D) has been obtained to date that reacts with this region of gp120. Although 450-D recognizes a linear epitope (amino acids 487–509) that is conserved between MN, SF2 and IIIB strains of HIV, it fails to show any neutralizing capabilities [39]. This confirms previous studies with polyclonal antibodies that the immunodominant region of gp120 at its C terminus does not induce neutralizing antibodies [43].

Epitopes on gp41

The other glycoprotein of HIV that plays an important role in HIV infectivity is the transmembrane glycoprotein, gp41. It serves as an anchor for gp120, with which it is associated by virtue of noncovalent interactions in several regions. In contrast to gp120, gp41 is a well-conserved molecule and shows no regions of hypervariability. The gp41 molecule is made up of approximately 345 amino acids, and its immunogenic regions have been mapped by several groups by using human sera and synthetic gp41-derived peptides.

Based on conserved hydrophilic sequences obtained by comparison of several different HIV strains, Gnann et al. [44] used peptides 8–26 amino acids long to define the epitopes of gp41 recognized by sera from infected individuals. Their study identified two overlapping regions, amino acids 579–604 and 593–604, derived from a highly conserved region in the extracellular domain of gp41 to be highly immunogenic in humans. A portion of the 593–604 epitope was homologous to the 21-amino-acid peptide SM284 previously reported to be immunodominant in individuals infected with HTLV-IIIB [45]. The strong immune reactivity against this region in humans was subsequently confirmed by several groups [46, 47]. Two additional immunogenic regions were described [21] by use of longer peptides (19–36 amino acids) derived from the gp41 sequence of HIV-1 and a pool of HIV-1-positive sera. These encompassed amino acids 771–802 and 845–862. The latter region had been reported by Klasse et al. [47] to be poorly antigenic, and antibodies to peptides 848–863 and 854–863, when detected, were present at low levels in humans. Klasse et al. [47] also described amino acids 654–666 as a weakly antigenic region.

Several laboratories have succeeded in making human MoAbs to the gp41 molecule (table 4) in an effort to create reagents that permit the differentiation between protective, inconsequential and deleterious components of the human immune response to gp41. Using EBV transformation of sensitized spleen cells from an HIV-seropositive individual, a MoAb IB8-env, directed against a well-conserved epitope of gp41 (amino acids 594–605) was obtained [48]. This epitope had previously been reported to be recognized by the vast majority of infected individuals. IB8env recognizes gp41 and gp160 in Western blots, indicating that it is not conformation-dependent. The MoAb IB8env neither neutralized HIV nor participated in ADCC.

Several anti-gp41 human MoAbs have been obtained by our group using EBV transformation of PBMCs from seropositive individuals [49, 50]. The antibodies were selected for their ability to bind to gp41 in whole-virus lysates. Epitope mapping for ten of these MoAbs has been done by using the recombinant peptide p121 which includes part of the extracellular domain of gp41 and spans amino acids 560–641 but is nonglycosylated, and a synthetic cyclic peptide spanning amino acids 579–613 that contains a disulfide bridge between the cysteines at amino acids 598 and 604. Hexapeptides overlapping by five amino acids were used for fine mapping of the epitopes.

Based on p121 binding and on competitive inhibition for reaction with HIV lysate, the ten anti-gp41 MoAbs could be divided into two distinct groups. The first group was comprised of MoAbs 50-69, 98-43, 181-D, 240-D

Table 4. Human MoAbs to gp41

MoAb	Specificity	Neutral-ization	ADCC	ADE	Reference
IB8env	GIWGCSGKLIC	–	–		Banapour et al. [48]
50-69	579–613	–	+	+	Gorny et al. [49], Xu et al. [50]
98-43	579–604	–	+		Robinson et al. [60]
181-D	qLLGIWg	–			Tyler et al. [9]
240-D	11gIWGcsg	–		+	
246-D	qqLLGIwg	–		+	
98-6	Conformational (Cluster II)	–	+		
120-16	644–663	–	+	+	
126-50	Conformational (Cluster II)	–	+		
167-7	Conformational (Cluster II)	–			
126-6	Conformational (Cluster II)	–			
41-7	N-C SGKLIC-C	–	–		Bugge et al. [51]
K14	643–692	–	–		Teeuwsen et al. [52]
CB-HIV-1	–				Grunow et al. [53]
M023	632–646	–	–		Ohlin et al. [54]
M030	677–681	–	–		
M043	687–691	–	–		
3IAI	gp41 peptide	–			Pollock et al. [55]
39A64	P (21) 465–641	–			
39B86	p24 peptide (pg2) 211–363	–			
V10-9	pENV9	–		+	Sugano et al. [56]
N2-4	pENV9	–		–	
13	pENV9	–		–	
NG3B7	467–531	–			Prigent et al. [57]
DB4B7	467–531	+			
2F5	–	+			Buchacher et al. [58]
4E10	–	+			

and 246-D, all of which bind p121 and display intragroup inhibition for reactivity with HIV lysate. All except 50-69 bound to a peptide spanning amino acids 579–604 and are therefore specific for linear sequences within this region which is located near the extracellular disulfide loop of gp41. Several studies with human sera had earlier shown this to be the immunodominant region of gp41 (see above). The fine specificity of three of these MoAbs was determined using hexapeptides. MoAb 181-D is directed towards an epitope consisting of qLLGIWg, 240-D towards 11glWGcsg and 246-D towards qqLLGIwg. (In each of these epitopes, the core of the epitope is represented in capital letters and the flanking amino acids by lower-case letters.) The dissociation constants of these MoAbs were studied both with the amino acid 579–604 (26-mer) peptide and the amino acid 560–641 (82-mer) peptide. In each case, the MoAbs had a higher affinity for the longer peptide, indicating that amino acids flanking the core epitope contribute to better maintenance of the native conformation and that the affinity for the native molecule would probably be higher than that observed with these peptides [50]. We define the epitopes recognized by this group of MoAbs as epitope cluster I. The fifth antibody of this group, MoAb 50-69, reacts with the cyclic peptide 579–613 in its unreduced form but not in the reduced form. Nor does 50-69 react with the peptide 579–604, which lacks the disulfide bond. These data indicate that MoAb 50-69 recognizes a conformational epitope dependent on the bond between the cysteines at positions 598 and 604.

The second group of anti-gp41 MoAbs comprises MoAbs 98-6, 120-16, 126-50, 126-6 and 167-7, none of which binds to p121, but all show intragroup inhibition in reacting with HIV lysate. MoAb 120-16 recognizes a linear epitope since it binds to peptide 644–663, whilst the others are directed towards conformational or discontinuous epitopes within this region, designated as epitope cluster II.

Studies directed at estimating the relative amounts of anti-cluster I and -cluster II antibodies in patients' sera were performed by measuring levels of serum antibodies that compete with MoAbs 50-69 (to cluster I) and 120-16 (to cluster II). Results show that cluster I dominates the immune response to gp41 in humans, eliciting almost 100 times more antibodies than does cluster II. This correlates well with data obtained with human sera and peptides wherein antibodies to the cluster I region have been reported consistently and at higher levels [21, 44, 46, 59] compared with antibodies to cluster II, which are found in fewer patients and at lower levels [21, 47, 59].

None of these anti-gp41 MoAbs are able to neutralize viral infectivity although all tested MoAbs (120-16, 126-50, 98-6, 50-69 and 98-43) are able to mediate ADCC [9], while only 120-16, 50-69, 240-D and 246-D can participate in complement-mediated, antibody-dependent enhancement of viral infection [17, 60]. Thus, the antibodies elicited by both of these regions of gp41 have apparent protective and deleterious biological functions.

Several other groups have also produced human MoAbs to gp41. Direct fusion of lymph node cells from an HIV-infected individual with a B lymphoblastoid cell line led to the production of an anti-gp41 MoAb 41-7 [51]. Epitope mapping of 41–7 by using overlapping peptides revealed that it was directed against amino acids 605–611 which lies within the immunodominant cluster I region described above. As with other antibodies to gp41, 41-7 has no virus-neutralizing capability.

Another human MoAb produced from transformed PBMCs of a seropositive individual, CB-HIV-1 gp41, binds to peptide 595–620 and 600–625 [53, 61]. These two peptides were also used successfully to differentiate sera from HIV-infected individuals and healthy controls, since essentially all infected subjects react to this region [61].

Another human MoAb to the cluster II epitope region of gp41 (amino acids 644–663), besides those described above, is K14, obtained by Teeuwsen et al. [52] from the PBMCs of an asymptomatic seropositive individual by EBV transformation. This MoAb recognizes an *E. coli*-recombinant env polypeptide containing amino acids 483–692 but not polypeptide 483–642. This indicates that the epitope of K14 lies in the region of amino acids 643–692. However, precise mapping by use of overlapping synthetic nonpeptides could not be achieved, apparently because this MoAb recognizes a conformational epitope, much like four of the five MoAbs to cluster II described above.

Prigent et al. [57] reported obtaining two anti-gp41 human MoAbs, again by EBV transformation of PBMCs from a seropositive individual. These MoAbs, NG3B7 and DB4B7, could react both with gp160 and gp41 from cell extracts, but not with HIV complexed to CD4-positive cells. One of these MoAbs, DB4B7, could neutralize HIV at 30 μg/ml in an assay involving inhibition of cytopathic activity of HIV. The MoAbs could also react with two recombinant proteins (61–531 and 467–758) which overlap in the 467–531 region which contains the splice site between gp120 and gp41, but the exact epitopes were not assessed.

Three additional human MoAbs, V10-9, N2-4 and 13, were described [17, 56], all of which were reactive with pENV9 [5], a recombinant protein containing amino acids 474–758 of HIV. Competitive binding studies suggested that these MoAbs were reactive with immunodominant epitopes of gp41 to which all infected individuals responded (cluster I). None of the MoAbs could neutralize HIV-1, and V10-9 enhanced HIV-1 infection in the presence of complement.

Attempts to recognize epitopes important for immune responsiveness of healthy, uninfected individuals have been made by in vitro immunization of PBMCs or spleen cells from healthy controls with HIV proteins [54, 55]. PBMCs were immunized using pENV9 (amino acids 474–758) and subsequently EBV-transformed. Three MoAbs, M023, M030 and M043, all IgM, were obtained, all of which reacted with pENV9 and were inhibited by peptides representing amino acids 632–646, 677–681 and 687–691, respectively, indicating that their epitopes were located in or near the transmembrane region (684–706) of gp41. None of these MoAbs could neutralize the virus or mediate ADCC.

In vitro immunization of normal human spleen cells with denatured HIV-1 followed by EBV transformation yielded three IgM MoAbs that curiously reacted with both the p24 and gp41 proteins of HIV. Epitope mapping of these antibodies could not be done, and none of them had any virus neutralization capability [55].

Buchacher et al. [58] have recently described two human hybridomas, 25F and 4E10, that make anti-gp41 MoAbs which show neutralization of HIV. Although the exact epitopes on gp41 recognized by these MoAbs are not known, the fusogenic region is not involved. Further characterization of these antibodies is awaited. The fact that, except for the two latter MoAbs to gp41, no anti-gp41 human MoAbs have shown neutralizing capability reflects on the need for screening with forms of gp41 that retain a structure and conformation close to the native conformation presented in vivo during infections.

Concluding Remarks

There are several human MoAbs available to begin to design immunoprophylactic reagents and immunotherapeutics for use in infected individuals. Although many MoAbs exist that are directed against the linear epitopes in the V3 region of gp120, there are also many antibodies to discontinuous conformational epitopes of gp120 that cannot be identified by peptide studies. This serves to emphasize the importance of continued efforts using diverse reagents and procedures for screening and production in order to obtain a larger repertoire of antibodies. These efforts are especially important in view of the emerging data on antigenic variation and neutralization-resistant mutants of HIV; such mutations that enable escape from neutralization have been reported in both gp120 and gp41 molecules [62–64]. Identification of more epitopes and a better understanding of the responses that they generate are urgently required.

The present technology is tedious and inefficient, making it imperative that quicker and easier methods be devised for obtaining human MoAbs. Attempts in this direction have been reported [65–67] and a panel of Fab fragments against gp120 have been generated by the use of recombinant DNA technology [68]. A combination of the immunological and recombinant DNA techniques should enable even more rapid progress in identification of epitopes of HIV.

Meanwhile, a multitude of broadly reactive human anti-HIV MoAbs have already been identified which are capable of neutralizing HIV-1 at nanogram or microgram levels in vitro. These reagents should be safe, nonimmunogenic and have half-lives in humans similar to those of human immunoglobulins. With these characteristics, these MoAbs represent reagents already available for the initiation of clinical trials. Used alone or in combination [69], they may serve to prevent transmission of HIV-1 and/or to prolong the lives of those already infected.

Acknowledgements

We wish to thank Carolyn Moore and Sara Leiman for their work in editing and preparing this manuscript, and Drs. Sylwia Karwowska and Miroslaw Gorny for their suggestions. We acknowledge the support of this work by the Research Center for AIDS and HIV Infection of the New York Veterans Affairs Medical Center, the NYU Center for AIDS Research (AI 27742), AIDS Training Grant (AI 07382), research grants from the NIH (AI 72658, AI 32424) and research funds from the Department of Veterans Affairs.

References

1 Javaherian K, Langlois AJ, LaRosa GJ, Profy AT, Bolognesi DP, Herlihy WC, Putney SD, Matthews TJ: Broadly neutralizing antibodies elicited by the hypervariable neutralizing determinant of HIV-1. Science 1990;250:1590.
2 Gorny MK, Xu JY, Gianakakos V, Karwowska S, Williams C, Sheppard HW, Hanson CV, Zolla-Pazner S: Production of site-selected neutralizing human monoclonal antibodies against the third variable domain of the HIV-1 envelope glycoprotein. Proc Natl Acad Sci USA 1991;88:3238.

3 Nara PL, Robey GW, Pyle SW, Hatch WC, Dunlop NM, Bess JW, Kelliher JC, Arthur LO, Fischinger PJ: Purified envelope glycoproteins from HIV-1 variants induce double type-specific neutralizing antibodies. J Virol 1988;62:2622.
4 Robey WG, Arthur LO, Matthews TJ, Langlois A, Copeland TD, Lerche NW, Oroszlan S, Bolognesi DP, Gilden RV, Fischinger PJ: Prospect for prevention of human immunodeficiency virus infection: Purified 120 kDa envelope glycoprotein induces neutralizing antibody. Proc Natl Acad Sci USA 1986;83:7023.
5 Putney SD, Matthews TJ, Robey G, Lynn DL, Robert-Guroff M, et al: HTLV-III/LAV-neutralizing antibodies to an *E. coli*-produced fragment of the virus envelope. Science 1986;234:1392.
6 Weiss RA, Clapham PR, Cheingsong-Popov R, Dalgleish AG, Carne CA, et al: Neutralization of human T-lymphotropic virus type III by sera of AIDS and AIDS-risk patients. Nature 1985;316:69.
7 Rusche JR, Javaherian K, McDanal C, Petro J, Lynn DL, Grimaila R, Langlois A, Gallo RC, Arthur LO, Fischinger PJ, Bolognesi DP, Putney SD, Matthews TJ: Antibodies that inhibit fusion of human immunodeficiency virus-infected cells bind a 24-amino acid sequence of the viral envelope, gp120. Proc Natl Acad Sci USA 1988;85:3198.
8 Tyler DS, Lyerly HK, Weinhold KJ: Minireview - Anti-HIV-1 ADCC. AIDS Res Hum Retroviruses 1989;5:557.
9 Tyler DS, Stanley SD, Zolla-Pazner S, Gorny MK, Shadduck P, Langlois AJ, Matthews TJ, Bolognesi DP, Palker TJ, Weinhold KJ: Identification of sites within gp41 that serve as targets for antibody-dependent cellular cytotoxicity by using human monoclonal antibodies. J Immunol 1990;145:3276.
10 Robinson WE Jr, Mitchell WM: Neutralization and enhancement of in vitro and in vivo HIV and simian immunodeficiency virus infections. AIDS 1990;4:S151.
11 Golding H, Shearer GM, Hillman K, Lucas P, Manischewitz J, Zajac RA, Clerici M, Gress RE, Boswell RN, Golding B: Common epitope in human immunodeficiency virus (HIV) I-gp41 and HLA class II elicits immunosuppressive autoantibodies capable of contributing to immune dysfunction in HIV 1-infected individuals. J Clin Invest 1983;83:1430.
12 Robinson J, Edward W, Michell WM, Chambers WH, Schuffman SS, Montefiori DC, Oeltmann TN: Natural killer cell infection and inactivation in vitro by the human immunodeficiency virus. Hum Pathol 1988;19:535.
13 Gras G, Strub T, Dormont D: Antibody-dependent enhancement of HIV infection. Lancet 1988;i:1285.
14 Homsy J, Meyer M, Tateno M, Clarkson S, Levy JA: The Fc and not CD4 receptor mediates antibody enhancement of HIV infection in human cells. Science 1989; 244:1357.
15 Robinson WE Jr, Montefiori DC, Mitchell WM: Antibody-dependent enhancement of human immunodeficiency virus type 1 infection. Lancet 1988;i:790.
16 Robinson WE, Montefiori DC, Mitchell WM, Prince AM, Alter HJ, Dreesman GR, Eichberg JW: Antibody-dependent enhancement of human immunodeficiency virus type 1 (HIV-1) infection in vitro by serum from HIV-1-infected and passively immunized chimpanzees. Proc Natl Acad Sci USA 1989;86:4710.
17 Robinson WE Jr, Kawamura T, Gorny MK, Montefiori DC, Mitchell WM, Lake D, Xu J, Matsumoto Y, Sugano T, Masuho Y, Hersh E, Zolla-Pazner S: Human monoclonal antibodies to the human immunodeficiency virus type 1 (HIV-1)

transmembrane glycoprotein gp41 enhance HIV-1 infection in vitro. Proc Natl Acad Sci USA 1990;87:3185.

18 Modrow S, Hahn BH, Shaw GM, Gallo RC, Wong-Staal F, Wolf H: Computer-assisted analysis of envelope protein sequences of seven human immunodeficiency virus isolates: Prediction of antigenic epitopes in conserved and variable regions. J Virol 1987;61:570.

19 Goudsmit J: Immunodominant B-cell epitopes of the HIV-1 envelope recognized by infected and immunized hosts. AIDS 1988;2:S41.

20 Palker TJ, Clark ME, Langlois AJ, Matthews TJ, Weinhold KJ, Randall RR, Bolognesi DP, Haynes G: Type-specific neutralization of HIV with antibodies to env-encoded synthetic peptides. Proc Natl Acad Sci USA 1988;85:1758.

21 Neurath AR, Strick N, Lee ESY: B cell epitope mapping of human immunodeficiency virus envelope glycoproteins with long (19- to 36-residue) synthetic peptides. J Gen Virol 1990;71:85.

22 Myers G, Korber B, Berzofsky JA, Smith RF, Pavlakis GN: Human retroviruses and AIDS. Published by Theoretical Biology and Biophysics Group T10, Los Alamos National Laboratory, Los Alamos, N.Mex., 1991.

23 Zwart G, Langeduk H, van der Hoek L, de Jong J, Wolfs TFW, Ramautarsing C, Bakker M, De Ronda A, Goudsmit J: Immunodominance and antigenic variation of the principal neutralization domain of HIV-1. Virology 1991;181:481.

24 LaRosa GJ, Davide JP, Weinhold K, Waterbury JA, Profy AT, Lewis JA, Langlois AJ, Dreesman GR, Boswell RN, Shadduck P, Holley LH, Karplus M, Bolognesi DP, Matthews TJ, Emini EA, Putney SD: Conserved sequence and structural elements in HIV-1 principal neutralizing determinant. Science 1990;249:932.

25 Javaherian K, Langlois AJ, McDanal C, Ross KL, Eckler LI, Jellis CL, Profy AT, Rusche JR, Bolognesi DP, Putney SD, Matthews TJ: Principal neutralizing domain of the human immunodeficiency virus type 1 envelope protein. Proc Natl Acad Sci USA 1989;86:6768.

26 LaRosa GJ, Weinhold K, Profy AT, Langlois AJ, Dreesman GR, Boswell RN, Shadduck P, Bolognesi DP, Matthews TJ, Emini EA, Putney SD: Conserved sequence and structural elements in the HIV-1 principal neutralizing determinant: Further clarification. Science 1991;253:1146.

27 Scott CF Jr, Silver S, Profy AT, Putney SD, Langlois A, Weinhold K, Robinson JE: Human monoclonal antibody that recognizes the V3 region of HIV gp120 and neutralizes the human T-lymphotropic virus type III-MN strain. Proc Natl Acad Sci USA 1990;87:8597.

28 Tilley JA, Honnen WJ, Warrier S, Racho ME, Chou T, Girard M, Muchmore E, Hilgartner M, Ho DD, Fung MSC, Pinter A: Potent neutralization of HIV-1 by human and chimpanzee monoclonal antibodies directed against three distinct epitope clusters of gp120; in Girard M, Valette L (eds): Retroviruses of Human AIDS and Related Diseases. 6ème Colloque des Cent Guardes, Marnes-la-Coquette, Paris, 1991, p 211.

29 Teng NN, Lam KS, Riera FC, Kaplan HS: Construction and testing of mouse-human heteromyelomas for human monoclonal antibody production. Proc Natl Acad Sci USA 1983;80:7308.

30 Karwowska S, Gorny MK, Buchbinder A, Zolla-Pazner S: 'Type-specific' human monoclonal antibodies cross-react with the V3 loop of various HIV-1 isolates; in

Brown F, Chanock RM, Ginsberg HS, Lerner RA (eds): Vaccines 92. Cold Spring Harbour Press, 1992, p 171.
31 Robinson JE, Holton D, Pacheco-Morell S, Liu J, McMurdo H: Identification of conserved and variant epitopes of human immunodeficiency virus type 1 (HIV-1) gp120 by human monoclonal antibodies produced by EBV-transformed cell lines. AIDS Res Hum Retroviruses 1990;6:567.
32 Putkonen P, Thorstensson R, Ghavamzadeh L, Albert J, Hild K, Biberfeld G, Norrby E: Prevention of HIV-2 and SIVsm infection by passive immunization in cynomolgus monkeys. Nature 1991;352:436.
33 Weiss RA, Clapham PR, Weber JN, Dalgleish AS, Lasky LA, Berman PW: Variable and conserved neutralization antigens of HIV. Nature 1986;324:572.
34 Ho DD, McKeating JA, Li XL, Moudgil T, Daar ES, Sun NC, Robinson JE: A conformational epitope on gp120 important in CD4 binding and HIV-1 neutralization identified by a human monoclonal antibody. J Virol 1991;65:489.
35 Ardman B, Kowalski M, Bristol J, Haseltine W, Sodroski J: Effects of CD4 binding of anti-peptide sera to the fourth and fifth conserved domains of HIV-1 gp120. J AIDS 1990;3:206.
36 Olshevsky U, Helseth E, Furman C, Li J, Haseltine W, Sodroski J: Identification of individual human immunodeficiency virus type 1 gp120 amino acids important for CD4 receptor binding. J Virol 1990;64:5701.
37 Tilley SA, Honnen WJ, Racho ME, Hilgartner M, Pinter A: A human monoclonal antibody against the CD4-binding site of HIV1 gp120 exhibits potent, broadly neutralizing activity. Res Virol 1991;142:247.
38 Posner MR, Hideshima T, Cannon T, Mukerjee M, Mayer KH, Byrn R: Human monoclonal antibody that reacts with HIV-1/pg120, inhibits virus binding to cells and neutralizes infection. J Immunol 1991;146:4325.
39 Karwowska S, Gorny MK, Buchbinder A, Gianakakos V, Williams C, Fuerst T, Zolla-Pazner S: Production of human monoclonal antibodies specific for conformational and linear non-V3 epitopes of gp120. AIDS Res Hum Retroviruses, in press.
40 Lake DF, Kawamura T, Tomiyama T, Robinson WE Jr, Matsumoto Y, Masuho Y, Hersh EM: Generation and characterization of a human monoclonal antibody that neutralizes diverse HIV-1 isolates in vitro. AIDS 1992;6:17.
41 Gorny MK, Conley AJ, Karwowska S, Buchbinder A, Xu J, Emini EA, Koenig S, Zolla-Pazner S: A human monoclonal antibody to the HIV-1 V3 envelope domain possesses broad neutralizing capacity. J Virol, in press.
42 McKeating JA, Thali M, Furman C, Karwowska S, Gorny MK, Cordell J, Zolla-Pazner S, Sodroski J, Weiss RA: Amino acid residues of the human immunodeficiency virus type 1 gp120 critical for the binding of rat and human neutralizing antibodies that block the gp120-sCD4 interaction. Submitted.
43 Palker TJ, Matthews TJ, Clark ME, Cianciolo GJ, Randall RR, Langlois AJ, White GC, Safai B, Snyderman R, Bolognesi DP, Haynes BF: A conserved region at the COOH terminus of human immunodeficiency virus gp120 envelope protein contains an immunodominant epitope. Proc Natl Acad Sci USA 1987;84:2479.
44 Gnann JW, Schwimmbeck PL, Nelson JA, Truax B, Oldstone MBA: Diagnosis of AIDS by using a 12-amino acid peptide representing an immunodominant epitope of the human immunodeficiency virus. J Infect Dis 1987;156:261.
45 Wang JJG, Steel S, Wisniewolski R, Wang CY: Detection of antibodies to human T-lymphotropic virus type III by using a synthetic peptide of 21 amino acid resi-

dues corresponding to a highly antigenic segment of gp41 envelope protein. Proc Natl Acad Sci USA 1986;83:1659.
46 Schrier RD, Gnann JW Jr, Langlois AJ, Shriver K, Nelson JA, Oldstone MBA: B- and T-lymphocyte responses to an immunodominant epitope of human immunodeficiency virus. J Virol 1988;62:2531.
47 Klasse PJ, Pipkorn R, Blomberg J: Presence of antibodies to a putatively immunosuppressive part of human immunodeficiency virus (HIV) envelope glycoprotein gp41 is strongly associated with health among HIV-positive subjects. Proc Natl Acad Sci USA 1988;85:5225.
48 Banapour B, Rosenthal K, Rabin L, Sharma V, Young L, Fernandez J, Engleman E, McGrath M, Reyes G, Lifson J: Characterization and epitope mapping of a human monoclonal antibody reactive with the envelope glycoprotein of human immunodeficiency virus. J Immunol 1987;139:4027.
49 Gorny MK, Gianakakos V, Sharpe S, Zolla-Pazner S: Generation of human monoclonal antibodies to HIV. Proc Natl Acad Sci USA 1989;86:1624.
50 Xu JY, Gorny MK, Palker T, Karwowska S, Zolla-Pazner S: Epitope mapping of two immunodominant domains of gp41, the transmembrane protein of human immunodeficiency virus type 1, using ten human monoclonal antibodies. J Virol 1991; 65:4832.
51 Bugge TH, Lindhardt BO, Hansen LL, Kusk P, Hulgaard E, Holmback K, Klasse PJ, Zeuthen J, Ulrich K: Analysis of a highly immunodominant epitope in the human immunodeficiency virus type 1 transmembrane glycoprotein, gp41, defined by a human monoclonal antibody. J Virol 1990;64:4123.
52 Teeuwsen VJP, Siebelink KHJ, Crush-Stanton S, Swerdlow B, Schalken JJ, Goudsmit J, van de Akker R, Stukart MJ, Uytdehaag FGCM, Osterhaus ADME: Production and characterization of a human monoclonal antibody reactive with a conserved epitope on gp41 of HIV-1. AIDS Res Hum Retroviruses 1990;6:381.
53 Grunow R, Jahn S, Porstmann T, Kiessig SS, Steinkellner SH, Steindl F, Mattanovich D, Gurtler L, Deinhardt F, Katinger H, von Baehr R: The high efficiency, human B cell immortalizing heteromyeloma CB-F7. J Immunol Methods 1988; 106:257.
54 Ohlin M, Broliden PA, Danielsson L, Wahren B, Rosen J, Jondal M, Borrebaeck CAK: Human monoclonal antibodies against a recombinant HIV envelope antigen produced by primary in vitro immunization: Characterization and epitope mapping. Immunology 1989;68:325.
55 Pollock BJ, McKenzie AS, Kemp BE, McPhee DA, D'Apice AJF: Human monoclonal antibodies to HIV-1: Cross-reactions with gag and env products. Clin Exp Immunol 1989;78:323.
56 Sugano T, Musuho Y, Matsumoto Y, Lake D, Gschwind C, Petersen EA, Hersh EM: Human monoclonal antibody against glycoproteins of human immunodeficiency virus. Biochem Biophys Res Commun 1988;155:1105.
57 Prigent S, Goossens D, Clerget-Raslain B, Bahraoui E, Roussel M, Tsiukas G, Laurent A, Montagnier L, Salmon C, Gluckman JC, Rouger P: Production and characterization of human monoclonal antibodies against core protein p25 and transmembrane glycoprotein gp41 of HIV-1. AIDS 1990;4:11.
58 Buchacher A, Predl R, Tauer C, Purtscher M, Gruber G, Heider R, Steindl F, Trkola A, Jungbauer A, Katinger H: Human monoclonal antibodies against gp41 and gp120 as potential agents for passive immunization; in Brown F, Chanock RM,

Ginsberg HS, Lerner RA (eds): Vaccines 92. Cold Spring Harbour Press, 1992, p 191.

59 Goudsmit J, Meloen RH, Brasseur R: Map of sequential B-cell epitopes of the HIV-1 transmembrane protein using human antibodies as probes. Intervirology 1990; 31:327.

60 Robinson WE Jr, Gorny MK, Xu JY, Mitchell WM, Zolla-Pazner S: Two immunodominant domains of gp41 bind antibodies which enhance HIV-1 infection in vitro. J Virol 1991;65:4169.

61 Dopel S, Porstmann T, Grunow R, Henklein P, Pas P, von Baehr R: Binding behaviour of antibodies reacting specifically with an immunodominant region of the transmembrane protein gp41 of HIV-1. J Virol 1990;28:189.

62 Reitz MS, Wilson C, Naugle C, Gallo RC, Robert-Guroff M: Generation of a neutralization-resistant variant of HIV-1 is due to selection for a point mutation in the envelope gene. Cell 1988;54:57.

63 Habeshaw JA, Dalgleish AG, Bountiff L, Newell AL, Wilks D, Walker LC, Manca F: AIDS pathogenesis: HIV envelope and its interaction with cell proteins. Immunol Today 1990;418:425.

64 Nara PL, Garrity RR, Goudsmit J: Neutralization of HIV-1: A paradox of humoral proportions. FASEB J 1991;5:2437.

65 Winter G, Milstein C: Man-made antibodies. Nature 1991;349:293.

66 Huse WD, Sastry L, Iverson SA, Kang AS, Alting-Mees M, Burton DR, Benkovic SJ, Lerner RA: Generation of a large combinatorial library of the immunoglobulin repertoire in phage lambda. Science 1989;246:1275.

67 Ward ES, Gussow D, Griffiths AD, Jones PT, Winter G: Binding activities of a repertoire of single immunoglobulin variable domains secreted from Escherichia coli. Nature 1989;341:544.

68 Burton DR, Barbas CF, Persson MA, Koenig S, Chanock RM, Lerner RA: A large array of human monoclonal antibodies to type 1 human immunodeficiency virus from combinatorial libraries of asymptomatic seropositive individuals. Proc Natl Acad Sci USA 1991;88:10134.

69 Buchbinder A, Karwowska S, Gorny MK, Burda ST, Zolla-Pazner S: Synergy between human monoclonal antibodies to HIV extends their effective biologic activity against homologous and divergent strains. AIDS Res Hum Retroviruses 1992;8:425.

Suman Laal, PhD, Research Center for AIDS and HIV Infection,
New York Veterans Affairs Medical Center, New York, NY 10010 (USA)

Norrby E (ed): Immunochemistry of AIDS.
Chem Immunol. Basel, Karger, 1993, vol 56, pp 112–126

Human Antibodies to HIV-1 by Recombinant DNA Methods

Dennis R. Burton, Carlos F. Barbas III

Departments of Immunology and Molecular Biology,
The Scripps Research Institute, La Jolla, Calif., USA

In 1989 the first experiments to generate monoclonal antibodies (MoAbs) by antigen selection from combinatorial libraries expressed in phage were described [1]. In these experiments, a mouse was immunized with the hapten nitrophenyl phosphonamidate and RNA was prepared from the spleen. After reverse transcription, the cDNAs of antibody heavy chains (Fd part of IgG1) and light chains were amplified by the polymerase chain reaction (PCR) and ligated into modified lambda phage vectors to give libraries of heavy and light chains. The two libraries were then 'crossed' by digestion of opposite arms of the vectors and religation to generate a random combinatorial library containing the genetic information for the production of Fab fragments. Screening of the library in a filter lift plaque assay using labeled hapten revealed a high frequency of positives which allowed the identification of 200 monoclonal Fab fragments. Analysis of 22 of these showed sequence diversity and apparent binding affinities of the order of $10^7 M^{-1}$.

The experiment was successful despite the scrambling of heavy and light chains inherent in the construction of random combinatorial libraries. One might have expected that this scrambling would render the chances of a productive heavy-light chain combination extremely improbable. However, two factors seem to work to generate a reasonable frequency of binders in the library [2]. One is the immune mRNA source which leads to a high representation in the library of chains arising from in vivo binders. The second is chain promiscuity, i.e. the ability of chains, particularly heavy, to accept more than one partner in productive antigen binding.

The methodology was then extended to allow generation of monoclonal Fab fragments from the peripheral-blood lymphocytes (PBLs) of an individ-

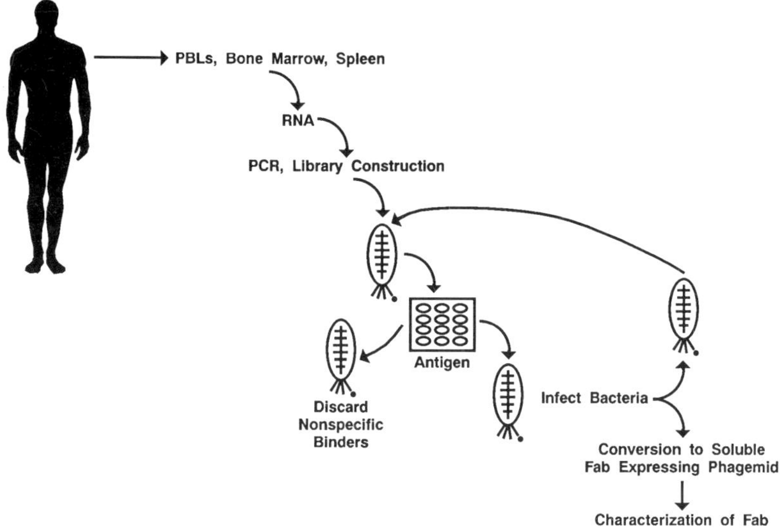

Fig. 1. Strategy for cloning human monoclonal Fab fragments from combinatorial libraries on the surface of phage.

ual recently boosted with tetanus toxoid [3]. The antibodies generated in this case showed considerable sequence diversity and apparent affinities in the range of 10^8–10^9 M^{-1}. Similar studies have been reported elsewhere [4]. We have also applied the method to the generation of antithyroglobulin antibodies from the thyroid tissue of an autoimmune individual [5].

These initial experiments with lambda phage clearly established the potential of combinatorial libraries to generate diverse specific high-affinity MoAbs [6]. However, the application of the lambda phage technology to the isolation of anti-gp120 antibodies from libraries derived from the PBLs or bone marrow cells of an HIV-1-seropositive individual failed despite extensive efforts. We do not know why but a reasonable explanation is the inherent 'stickiness' of gp120 complicating the filter lift assay. Meanwhile, we developed a new system involving the expression of random combinatorial libraries on the surface of M13 phage [7, 8]. Similar systems have been described for the expression of antibodies [9, 10] and other proteins [11] on the surface of phage based on the work of Smith [12].

The system can be used to generate MoAbs against gp120 as described below. The basic strategy is summarized in figure 1.

Phage Display of Antibody Combinatorial Libraries and Selection of Specific Fab Fragments

The heavy (Fd region) and light chains are cloned sequentially into the phagemid vector pComb3 [8, 13] (fig. 2). The vector is constructed to fuse Fd with the C-terminal domain of the minor coat protein III (cpIII). This fusion protein is targeted to the periplasmic space of *Escherichia coli* where it is anchored in the inner membrane by the cpIII domain. The light chain then assembles on the heavy chain template to give an Fab fragment. Inclusion of the F1 intergenic region in the vector means that superinfection with M13 helper phage leads to packaging of single-stranded phagemid, including heavy- and light-chain genes. Normal phage morphogenesis then leads to incorporation of the Fab-cpIII fusion and the native cpIII into the virion. Native cpIII is necessary for infection and the fusion, generally one copy per phage particle, is displayed for antigen selection. The pComb3 system thus links antibody recognition with the genes for antibody production.

Following construction of a combinatorial library on the surface of M13 phage using the pComb3 system, the phage is panned against antigen in ELISA wells. After vigorous washing, the bound phage, enriched for those bearing specific surface Fabs, is eluted with acid. This phage is then amplified followed by further rounds of panning. In this way one can rapidly generate a panel of antigen-specific Fabs. For instance, the first experiments began with a human combinatorial anti-tetanus toxoid library in which the frequency of positives was about 1 in 5,000 and this number was 1 in 4 after one round of panning against toxoid, 7 out of 10 following 2 rounds and 9 out of 10 following 3 rounds. In another study, a library was prepared which included known anti-tetanus toxoid clones at a frequency of about 1 in 170,000. Three rounds of panning against toxoid were found to give enrichment such that 20/20 clones were antigen-specific, indicating that the method could access clones of low abundance. The method also allows sorting of clones based on affinity as shown by a 250-fold enrichment of a tight binding tetanus toxoid clone (apparent K_a approximately 10^9 M^{-1}) relative to a weaker binding clone (10^7 M^{-1}) in a mixture of the two following one round of panning.

A further feature of the pComb3 system is that, once a surface display phagemid has been selected, the cpIII gene can be excised in a NheI/SpeI digestion and the vector religated via the compatible cohesive ends. The phagemid can then be used to express soluble Fab fragment in the conventional manner (fig. 3).

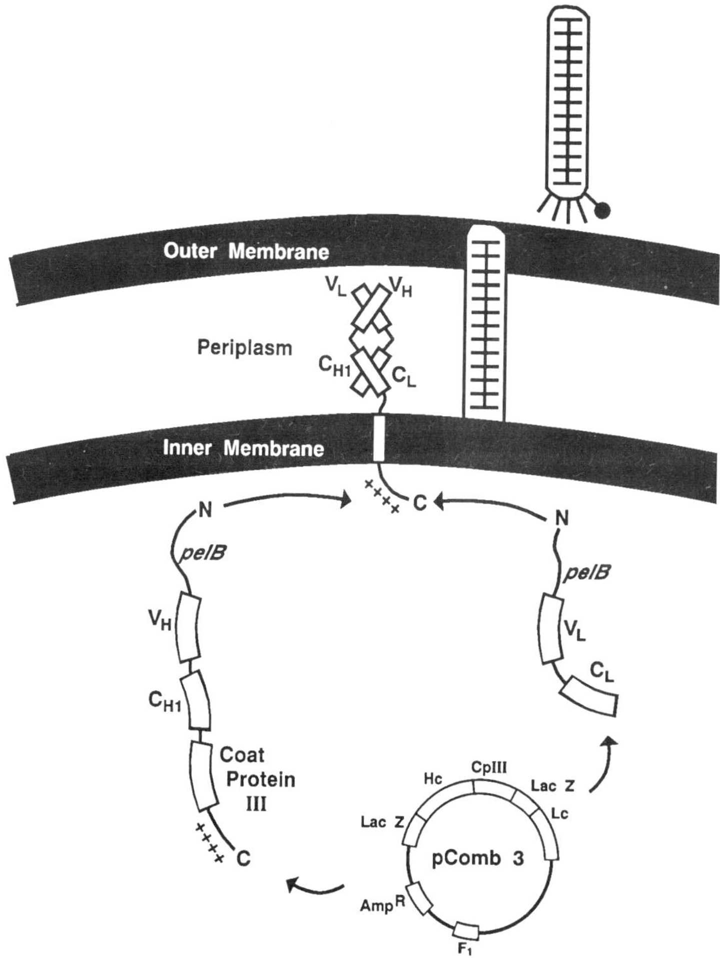

Fig. 2. Composition of the pComb3 vector and proposed pathway for Fab assembly and incorporation into the phage coat. Expression of Fd (Hc; heavy chain)/cpIII fusion and light chain (Lc) is controlled by lac promoter/operator sequences. The chains are directed to the periplasmic space by pelB signal sequences which are subsequently cleaved. The heavy chain is anchored to the membrane by the cpIII fusion whereas the light chain is secreted into the periplasm. The two chains then assemble on the membrane. The Fab molecule (●) is then incorporated into the phage coat during extrusion of the phage through the membrane via the cpIII segment. Reproduced from Barbas et al. [13] with permission.

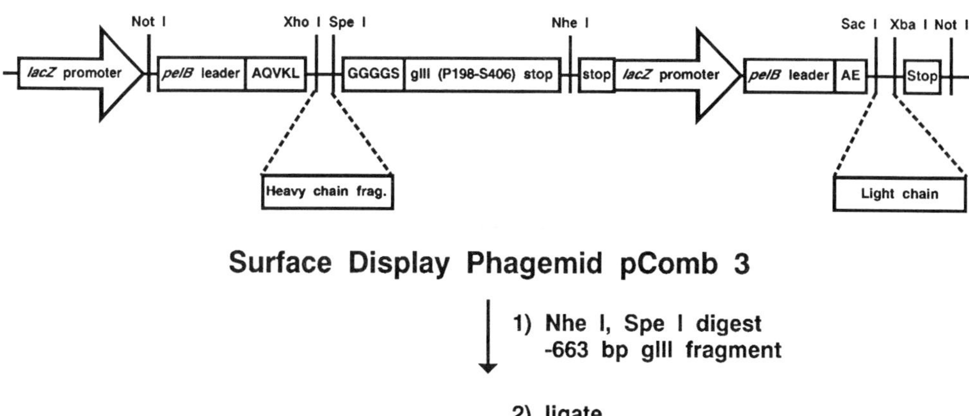

Fig. 3. Conversion of pComb3 to a system expressing soluble Fab. Reproduced from Barbas et al. [8] with permission.

Application of the Combinatorial Library Approach to the Generation of Antibodies to HIV-1

We have now applied the pComb3 system to the generation of anti-HIV antibodies [14]. The first library was prepared from a 31-year-old homosexual male who has been HIV-positive for 6 years but has no symptoms of disease. Serological studies showed the presence of a significant ELISA titer (1:3,000) against the HIV-1 surface glycoprotein, gp120 (IIIB strain). After securing informed consent, bone marrow cells were obtained by aspiration. Amplified antibody genes were then cloned into pComb3 to give a library of 10^7 members.

The phage surface expression library was panned against recombinant gp120 (strain IIIB) coated on ELISA wells. Four rounds of panning produced an amplification in eluted phage of a factor of about 100, indicating enrichment for specific antigen-binding clones. Forty reconstructed clones secreting soluble Fab fragments were grown and the supernates screened in an ELISA assay for reactivity with recombinant gp120. The supernates from

33 clones showed clear reactivity. The supernates did not react with BSA-coated wells and anti-tetanus toxoid Fab supernates did not react with gp120-coated wells.

DNA from the 33 clones was used as templates for sequencing of thymidine nucleotides of the variable-heavy (V_H) and variable-light (V_L) regions to reveal that at least 10 clones had unique heavy chains and 20 clones unique light chains. A representative number of chains were then sequenced. Table 1 indicates the diversity of the panel of antibodies cloned. The heavy chains fell into 6 families, as shown, with evidence of somatic mutation within a given family (compare clones b3 and b5). The light chains showed even greater diversity. Chain promiscuity was observed in the sense that a very similar or identical heavy chain was found paired with a different light chain, e.g. clones 3 and 5.

To measure the affinities of the Fab fragments for gp120, inhibition ELISAs using soluble gp120 were performed. The examination of 15 clones showed that most inhibition constants were less than 10^{-8} M, implying monomer Fab-gp120 apparent binding constants of the order of or greater than 10^8 M^{-1} (fig. 4).

The phage surface library was also panned against gp160 (IIIB), gp120 (SF2) and a constrained peptide having the central part of the MN V3 loop sequence. Fabs isolated by panning against gp160 and showing a strong ELISA signal with gp160 also cross-reacted strongly with gp120. The sequences of the Fabs obtained by panning against gp160 (IIIB) or gp120 (SF2) were mostly closely related to those described above from panning against gp120 (IIIB). Indeed, regardless of the panning antigen [gp120 (IIIB), gp120 (SF2) or gp160 (IIIB)], the Fabs selected reacted with both gp120 (IIIB) and gp120 (SF2). Several Fabs were obtained by panning against the constrained peptide, but only one reacted with gp120. In fact, this Fab reacted with gp120 (SF2) but not gp120 (IIIB).

The ability of Fabs and soluble CD4 to compete for gp120 was investigated in competition ELISAs. All of the Fabs obtained by panning against gp120/160 were found to be competed by CD4 for binding to gp120, a number of clones being illustrated in figure 5 [15]. The Fab binding to the V3 loop was not competed.

Therefore, the predominant Fabs isolated from this donor by the library approach appear to be strain-cross-reactive and CD4-inhibited. This is consistent with the observation that more than 50% of the reactivity of the donor serum with gp120 (IIIB) is inhibitable by CD4. Further, the Fabs appear to be directed to major epitopes on gp120, in that a cocktail of three of

Table 1. Variable domain sequences of Fabs binding to gp120

V_H domains

Clone	FR1	CDR1	FR2	CDR2
b1	LEESGTEFKPPGSSVKVSCKASGGTFG	DYASNYAIS	WVRQAPGQGLEYIG	GITPTSGSADYAQKFQG
b3	LEESGGRLVKPGGSLRLSCEGSGFTFT	NAWMT	WVRQSPGKGLEWVA	SIKSKFDGGSPHYAAPVEG
b5	LEQSGGGLVKPGGSLRLSCEGSGFTFT	NAWMT	WVRQSPGKGLEWVA	SIKSKFDGGSPHYAAPVEG
b4	LEQSGAEVKKPGASVKVSCQASGYRFS	NFVIH	WVRQAPGQRFEWMG	WINPYNGNKEFSAKFQD
b6	LEESGGGLVKPGGSLRLSCVGSGFTFS	SAWMA	WVRQAPGRGLEWVG	LIKSKADGETTDYATPVKG
b8	LEESGEAVVQPGRSLRLSCAASGFIFR	NYAMH	WVRQAPGKGLEWVA	LIKYDGRNKYYADSVKG
b11	LEQSGGGVVKPGGSLRLSCEGSGFTFP	NAWMT	WVRQSPGKGLEWVA	SIKSKFDGGSPHYAAPVEG

Clone	FR3	CDR3	FR4	*J gene*
b1	RVTISADRFTPILYMELRSLRIEDTAIYYCAR	ERRERGWNPRALRGALDF	WGQGTRVFVSP	J_H3
b3	RFSISRNDLEDKMFLEMSGLKAEDTGVYYCAT	KYPRYSDMVTGVRNHFYMDV	WGKGTTVIVSS	J_H6
b5	RFTISRNDLEDKLFLEMSGLKAEDTGVYYCAT	KYPRYFDMMAGVRNHFYMDV	WGTGTTVIVSS	
b4	RVTFTADTSANTAYMELRSLRSADTAVYYCAR	VGPYSWDDSPQDNYYMDV	WGKGTTVIVSS	J_H6
b6	RFSISRNNLEDTVYLQMDSLRADDTAVYYCAT	QKPRYFDLLSGQYRRVAGAFDV	WGHGTTVTVSP	J_H3
b8	RFTISRDNSKNTLYLQMNSLRAEDTAVYYCAR	DIGLKGEHYDILTAYGPDY	WGQGTLVTVSS	J_H4
b11	RFTISRNDLEDKVFLQMNGLKAEDTGVYYCAT	RYPRYSEMMGGVRKHFYMDV	WGKGTTVSVSS	J_H6

V_κ domains

Clone	FR1	CDR1	FR2	CDR2
b1	ELTQSPSSLSASVGDRVTITC	RASQGISNYLA	WYQQKPGKVPRLLIY	AASTLQP
b3	ELTQSPGTLSLSPGERATLSC	RASHRVNNNFLA	WYQQKPGQAPRLLIS	GASTRAT
b5	ELTQSPASVSASVGDTVTITC	RASQDIHNWLA	WYQQQPGKAPKLLIY	AASSLQS
b4	ELTQSPGTLSLSPGERATFSC	RSSHSIRSRRVA	WYQHKPGQAPRLVIH	GVSNRAS
b6	ELTQSPGTLSLSPGERATLSC	RAGQSISSNYLA	WYQQKPGQAPRLLIY	GASNRAT
b8	ELTQSPSSLSASVGDRVTITC	RASQSISNYLN	WYQQKPGKAPKLLIY	AASSLQR
b11	ELTQSPGTLSLSPGERATLSC	RASQRVNSNYLA	WYQQKPGQTPRVVIY	STSRRAT

Clone	FR3	CDR3	FR4	*J gene*
b1	GVPSRFSGSGSGTDFTLTISSLQPEDVATYYC	QKYNSAPRT	FGQGTKVEIKRT	JK1
b3	GIPDRFSGSGSGTDFTLTISRLEPDDFAVYYC	QQYGDSPLYS	FGQGTKLEIKRT	JK2
b5	GVPSRFSGRGSGTDFTLTISSLQPEDFATYYC	QQGNSFPK	FGPGTVVDIKR	JK3
b4	GISDRFSGSGSGTDFTLTITRVEPEDFALYYC	QVYGASSYT	FGQGTKLERKRT	JK2
b6	GIPDRFSGSGSGTDFTLSISRLEPEDFAVYYC	QQYGTSPYT	FGQGTQLDIKRT	JK2
b8	GVPSRFSGSGSGTDFTLSISSLQPEDFATYYC	QQSYSIPPLT	FGGGTKVEIKRT	JK4
b11	GVPDRFSGSGSGTDFTLTISRLEPEDFAVYYC	QQFGDAQYT	FGQGTKLEIKRT	JK2

VH domains can be organized into 6 groups by sequence; clones b3 and b5 belong to the same group, the other clones are representative of their groups. The VL domains are constrained into the groupings defined by the VH domain sequences.

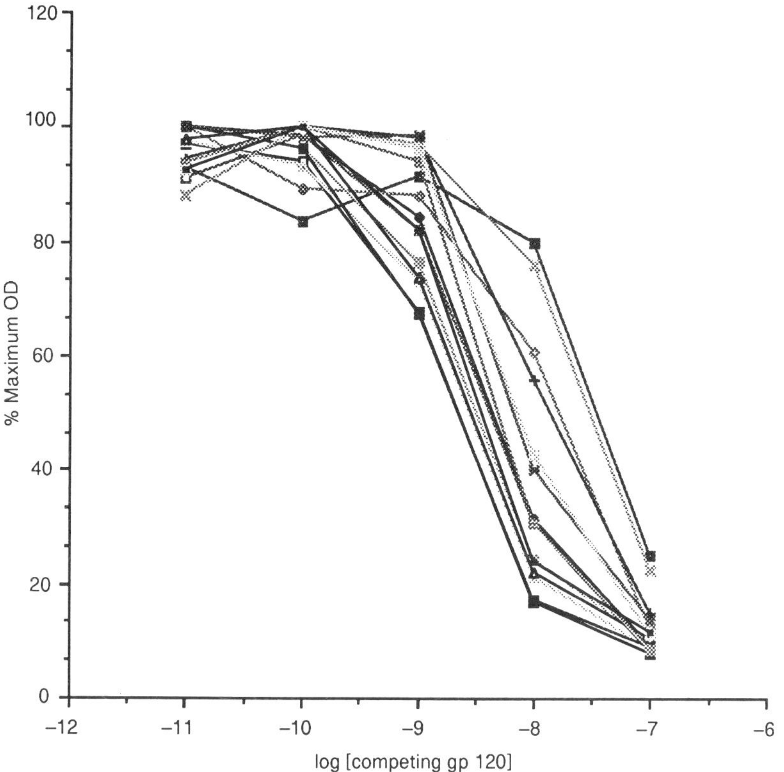

Fig. 4. Apparent affinities of Fabs for gp120 estimated by inhibition ELISA for 15 different clones. Reproduced from Burton et al. [14] with permission.

the Fabs was able to inhibit >50% of the serum reactivity with gp120 (IIIB) of more than 90% of a selection of seropositive donors [R. Burioni, pers. communication].

Neutralization of HIV-1 by Recombinant Fabs

We have begun to look at the neutralizing ability of the Fabs in collaboration with Dr. Peter Nara of the Laboratory of Tumor Biology, NCI, Frederick, Md., and Prof. Erling Norrby, Department of Virology, Karolinska Institute, Stockholm [16]. Neutralization was determined as the ability of Fabs to inhibit infection as measured in both p24 ELISA and

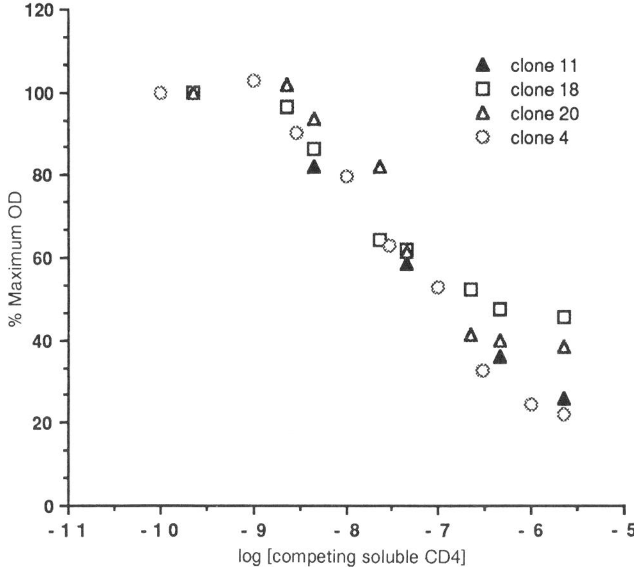

Fig. 5. Competition of CD4 and Fab fragments for gp120 in an ELISA assay. Reproduced from Barbas et al. [15] with permission.

syncytium assays. One group of closely sequence-related Fabs was found to neutralize virus in both assays with a titer (50% neutralization) at approximately 1 μg/ml. Another Fab neutralized in the p24 ELISA but not the syncytial assay. The majority of Fabs showed weak or no neutralizing ability. The results imply that virion aggregation or cross-linking of gp120 molecules on the virion surface is not an absolute requirement for HIV-1 neutralization. Further, the observation that all of the Fabs were competitive with soluble CD4 for binding to gp120 and yet few neutralize effectively implied that the mechanism of neutralization in this case may not involve receptor blocking.

Genetic Manipulation of Antibodies

Following identification of antigen-specific Fabs, the clones may be manipulated to improve their affinity, alter their specificity or express the entire antibody by addition of the Fc portion. One strategy which should prove to be of general use in altering the properties of antibodies isolated from combinatorial libraries is chain shuffling. Chain shuffling usually involves the defined cross of a given chain from an isolated clone with complementary unselected chains. Selection for binding from this constrained library can yield higher-affinity clones [T. Collet, pers. communication] and clones with changes in fine specificity [17]. Shuffling is a constructive approach since all possible H/L combinations may not be present in the original library and this focused search can reveal a spectrum of related clones. Generation of families of functionally related clones may be of use in therapy should anti-Id responses prove to be important in the repeated long-term administration of therapeutic antibodies.

Alternative strategies for the refinement of clones involve mutagenesis and reselection. This, in effect, attempts to carry the evolution of the antibody farther than the body has. Random mutagenesis of the variable regions is possible by a number of different approaches such as chemical mutagenesis [18], polymerase-induced mutagenesis [19] and in vivo mutagenesis using mutator strains of *E. coli* [20]. Focused mutagenesis, in which several residues are targeted, does not mimic the supposed random mutation and selection of the immune system, but allows all possible mutations in a defined region to be explored. This strategy has been successful in the generation of high-affinity variants of human growth hormone [21]. For antibodies, the strategy involves targeting specific CDR regions for mutagenesis since this is most likely to improve affinity and least likely to create problems of immunogenicity.

Although positive selection for variants of increased affinity is an obvious aim, mutagenesis could also be employed to increase or decrease cross-reactivity. For example, it may be desirable to produce antibodies that are highly cross-reactive in their binding to gp120 from a number of different strains of HIV. The selection strategy would then involve panning against alternate types of gp120 or a mixture. Specificity can conversely be increased by including the antigen giving rise to unwanted cross-reactivity in the wash solution during the selection or by pre-selection of the phage with the antigen. This effectively allows a negative selection to be performed.

Expression of whole antibody molecules, at least in part because of the glycosylation requirement of the Fc part of the molecule, demands the use of eukaryotic cell lines. The favorites in this regard have been myeloma [22] or Chinese hamster ovary (CHO) cells [23], although baculovirus has also been used [24]. To couple this technology to phage Fab technology, eukaryotic vectors need to be modified to accept the heavy and light chain products of the phage system. In principle any Fc can be linked to the Fd of the heavy chain from the phage system. We have shown how a whole IgG1κ molecule can be expressed in CHO cells using the Fd and κ-chains derived from phage [E. Bender, in preparation]. As expected, the molecule binds antigen with retained antigen affinity.

Gene Rescue from Antibody-Producing Cell Lines

Epstein-Barr virus transformation has been successfully used to generate human cell lines secreting MoAbs specific for HIV-1 proteins [see Zolla-Pazner, this volume]. Such lines can be unstable or low secretors of antibody. Alternatively, the antibody may not be of the isotype desired. In any of these cases, it may be appropriate to rescue heavy and light chain genes from mRNA from the cell line and express the Fab in bacteria. We have described this process for the cloning and expression of a human anti-rhesus D antibody [25]. In this case, 5/5 clones examined had the correct heavy and light chains. This is often not the case for hybridomas which appear to contain mRNA from other chains which is PCR-amplified and cloned. For instance, for one mouse hybridoma only 1 in 1,000 recombinants had antigen-binding activity [L. Sastry, pers. communication]. Therefore, the most prudent general strategy may be to clone from the cell line into a phage display vector such as pComb3 and then pan against antigen to select positive clones.

Alternatives to Seropositive Humans as Sources of Antibody Libraries

There are a number of alternative sources for antibody libraries which may prove complementary to the use of seropositive humans. These are naive libraries, synthetic or semisynthetic libraries, chimpanzees and severe combined immunodeficiency (SCID) mice populated with human cells. Preparation of naive libraries involves the use of RNA from nonimmune sources and amplification of μ- or δ-heavy chains which are the starting point in the natural response. Naive libraries have been constructed from both humans and mice and have been selected from with some success for the production of low- or medium-affinity antibodies [26, 27]. There are two likely pitfalls of antibodies selected from naive repertoires: relatively low affinity for the antigen and possible deleterious cross-reactivity. Improvements on both points would necessitate utilization of the strategies discussed above.

Synthetic or semisynthetic antibody libraries are the second alternative to an immune source. Initial explorations of semisynthetic antibodies utilized a single clone selected from a library from an immune patient [28]. A 16-amino-acid random sequence was then introduced over the CDR3 region of the heavy chain to generate a vast library of antibodies. Selection of the library against a variety of antigens allowed for the cloning of new specificities. The complete randomization of 16 amino acids would require the generation of a library of greater than 10^{20} clones, far in excess of the number which is obtainable by the transformation of *E. coli*. However, libraries can be constructed which match or exceed the diversity of clones examined by an animal at a given moment, approximately 10^7. Selection of this library derived from a human antitetanus clone against a new antigen, fluorescein, resulted in the isolation of clones which bound fluorescein with affinities approaching that obtained by the secondary boost of a mouse. This strategy is proving useful for the generation of antibodies against a variety of antigens. Extension of this strategy to the synthesis of all the CDRs or the use of natural libraries of FR1-FR3 fragments in combination with synthetic CDR3s should yield libraries from which almost any given specificity is retrievable. This approach may be limited by the same pitfalls as the naive approach. However, it has one distinct advantage. Diversity of these libraries is controlled at the level of nucleic acid synthesis, whereas the diversity of a naive library is limited by the source of RNA which is susceptible to bias by RNA derived from plasma cells or activated B cells.

Given that the extent of differences between chimpanzees and humans in constant domain sequences is similar to that between allotypes of humans [29, 30], an alternative source to seropositive humans is immunized chimpanzees. We have indeed found that the human PCR primers will successfully amplify $\gamma 1$ heavy and κ and λ light chains and it would thus appear that the library approach should be feasible.

SCID mice populated with human cells offer the possibility of antigen boosting of human responses outside the human body. We have shown that SCID mice can be used in conjunction with the combinatorial library approach [31]. Thus, we were able to populate a SCID mouse with PBLs from a donor who had not had contact with tetanus toxoid for more than 17 years, generate a secondary response in the mouse and clone antigen-specific high-affinity human Fab fragments. For HIV-seropositive donors this sort of approach might be useful in stimulating antibodies against HIV-1 which form only a small part of the current response or indeed those relegated to the memory compartment because of changes in the nature of the virus. One also has the advantage of being able to boost with smaller units than the virus. So, for instance, a response to a particular peptide could be stimulated and human monoclonal Fab fragments rescued. In the future, it may also become possible to create secondary responses from seronegative donors.

Finally, the distinct advantage of utilizing an immune source, man or chimpanzee, is that the antibodies have been both highly positively selected and negatively selected. These antibodies which are tried and tested in the animal, are likely to be the best source of therapeutic antibodies, though refinement by way of in vitro methods may prove important.

Acknowledgements

We would like to acknowledge the support and enthusiasm of Richard A. Lerner and the efforts and guidance of many valued collaborators and coworkers. C.F.B. is a Scholar of The American Foundation for AIDS Research.

References

1 Huse WD, Sastry L, Iverson SA, Kang AS, Alting-Mees M, Burton DR, Benkovic SJ, Lerner RA: Generation of a large combinatorial library of the immunoglobulin repertoire in phage lambda. Science 1989;246:1275–1281.
2 Kang AS, Jones TM, Burton DR: Antibody redesign by chain shuffling from random combinatorial immunoglobulin libraries. Proc Natl Acad Sci USA 1991;88:11120–11123.
3 Persson MAA, Caothien RH, Burton DR: Generation of diverse high-affinity human monoclonal antibodies by repertoire cloning. Proc Natl Acad Sci USA 1991;88: 2432–2436.
4 Mullinax RL, Gross EA, Amberg JF, Hay BN, Hogrefe HH, Kubitz MM, Greener A, Alting-Mees M, Ardourel D, Short JM, Sorge JA, Shopes B: Identification of human antibody fragment clones specific for tetanus toxoid in a bacteriophage lambda immunoexpression library. Proc Natl Acad Sci USA 1990;87:8095–8099.
5 Hexham JM, Furmaniak J, Persson MAA, Pegg C, Burton DR, Smith BR: Cloning and expression of a human thyroglobulin autoantibody. Autoimmunity 1991;11:69–70.
6 Burton DR: Human and mouse monoclonal antibodies by repertoire cloning. Trends Biotechnol 1991;9:169–175.
7 Kang AS, Barbas CF III, Janda KD, Benkovic SJ, Lerner RA: Linkage of recognition and replication functions by assembling combinatorial antibody Fab libraries along phage surfaces. Proc Natl Acad Sci USA 1991;88:4363–4366.
8 Barbas CF III, Kang AS, Lerner RA, Benkovic SJ: Assembly of combinatorial antibody libraries on phage surfaces. The gene III site. Proc Natl Acad Sci USA 1991;88: 7978–7982.

9 McCafferty J, Griffiths AD, Winter G, Chiswell DJ: Phage antibodies: Filamentous phage displaying antibody variable domains. Nature 1990;348:552–554.

10 Clackson T, Hoogenboom HR, Griffiths AD, Winter G: Making antibody fragments using phage display libraries. Nature 1991;352:624–628.

11 Bass S, Greene R, Wells JA: Hormone phage: An enrichment method for variant proteins with altered binding properties. Proteins Struct Funct Genet 1990;8:309–314.

12 Smith GP: Filamentous fusion phage. Novel expression vectors that display cloned antigens on the virion surface. Science 1985;228:1315–1316.

13 Barbas CF III, Lerner RA: Combinatorial immunoglobulin libraries on the surface of phage (Phabs): Rapid selection of antigen-specific Fabs; in Lerner RA, Burton DR (eds): Methods: A Companion to Methods in Enzymology. Orlando, Academic Press, 1991, vol 2, pp 119–124.

14 Burton DR, Barbas CF III, Persson MAA, Koenig S, Chanock RM, Lerner RA: A large array of human monoclonal antibodies to type 1 human immunodeficiency virus from combinatorial libraries of asymptomatic seropositive individuals. Proc Natl Acad Sci USA 1991;88:10134–10137.

15 Barbas CF III, Persson MAA, Koenig S, Chanock RM, Burton DR, Lerner RA: A large array of human monoclonal antibodies to HIV-1 from combinatorial libraries of an asymptomatic seropositive individual; in Brown F, Chanock RM, Ginsberg H, Lerner RA (eds): Vaccines '92, Modern Approaches to New Vaccines Including Prevention of AIDS. Cold Spring Harbor, Cold Spring Harbor Laboratory Press, 1992, pp 9–12.

16 Barbas CF III, Bjorling E, Chiodi F, Dunlop N, Cababa D, Jones TM, Zebedee SL, Persson MAA, Nara PL, Norrby E, Burton DR: Recombinant human Fab fragments neutralize human type 1 immunodeficiency virus in vitro. Proc Natl Acad Sci USA, in press.

17 Zebedee SL, Barbas CF III, Hom Y-L, Caothien RH, Graff R, DeGraw J, Pyati J, LaPolla R, Burton DR, Lerner RA, Thornton GB: Human combinatorial antibody libraries to hepatitis B surface antigen. Proc Natl Acad Sci USA 1992;89:3175–3179.

18 Myers RM, Lerman LS, Maniatis T: A general method for saturation mutagenesis of cloned DNA fragments. Science 1985;229:242–247.

19 Leung DW, Chen E, Goeddel DV: A method for random mutagenesis of a defined DNA segment using a modified polymerase chain reaction. Tech J Methods Cell Mol Biol 1989;1:11–15.

20 Schaaper RM: Mechanisms of mutagenesis in the *Escherichia coli* mutator *mutD5*: Role of DNA mismatch repair. Proc Natl Acad Sci USA 1988;85:8126–8130.

21 Lowman HB, Bass SH, Simpson N, Wells JA: Selecting high-affinity binding proteins by monovalent phage display. Am Chem Soc 1991;30:10832–10838.

22 Wright A, Shin S-U: Production of genetically engineered antibodies in myeloma cells: Design, expression and applications; in Lerner RA, Burton DR (eds): Methods: A Companion to Methods in Enzymology. Orlando, Academic Press, 1991, vol 2, pp 125–135.

23 Bebbington CR: Expression of antibody genes in nonlymphoid mammalian cells; in Lerner RA, Burton DR (eds): Methods: A Companion to Methods in Enzymology. Orlando, Academic Press, 1991, vol 2, pp 136–145.

24 Hasemann CA, Capra JD: Baculovirus expression of antibodies: A method for the expression of complete immunoglobulins in a eukaryotic host; in Lerner RA, Burton

DR (eds): Methods: A Companion to Methods in Enzymology. Orlando, Academic Press, 1991, vol 2, pp 146–158.

25 Williamson RA, Persson MAA, Burton DR: Expression of a human monoclonal anti-(rhesus D) Fab fragment in *Escherichia coli* with the use of bacteriophage lambda vectors. Biochem J 1991;277:561–563.

26 Marks JD, Hoogenboom HR, Bonnert TP, McCafferty J, Griffiths AD, Winter G: By-passing immunization human antibodies from V-gene libraries displayed on phage. J Mol Biol 1991;222:581–597.

27 Gram H, Marconi L-A, Barbas CF III, Collet TA, Lerner RA, Kang AS: In vitro selection and affinity maturation of antibodies from a naive combinatorial immunoglobulin library. Proc Natl Acad Sci USA 1992;89:3576–3580.

28 Barbas CF III, Bain JD, Hoekstra DM, Lerner RA: Semisynthetic combinatorial antibody libraries: A chemical solution to the diversity problem. Proc Natl Acad Sci USA 1992;89:4457–4461.

29 Ehrlich PH, Moustafa ZA, Harfeldt KE, Isaacson C, Ostberg L: Potential of primate monoclonal antibodies to substitute for human antibodies: Nucleotide sequence of chimpanzee Fab fragments. Hum Antibod Hybridomas 1990;1:23–26.

30 Ehrlich PH, Moustafa ZA, Osterberg L: Nucleotide sequence of chimpanzee F_c and hinge regions. Mol Immunol 1991;28:319–322.

31 Duchosal MA, Eming S, Fischer P, Leturcq D, Barbas CF III, McConahey PJ, Caothien RH, Thornton GB, Dixon FJ, Burton DR: Immunization of hu-PBL-SCID mice and the rescue of human monoclonal Fab fragments through combinatorial libraries. Nature 1992;355:258–262.

Dr. D.R. Burton, Department of Immunology, The Scripps Research Institute, 10666 North Torrey Pines Road, La Jolla, CA 92037 (USA)

Dr. C.F. Barbas, Department of Molecular Biology, The Scripps Research Institute, 10666 North Torrey Pines Road, La Jolla, CA 92037 (USA)

Norrby E (ed): Immunochemistry of AIDS.
Chem Immunol. Basel, Karger, 1993, vol 56, pp 127–149

CD4+ T Cell Epitopes in HIV-1 Proteins

Robert F. Siliciano

Department of Medicine, Johns Hopkins University School of Medicine, Baltimore, Md., USA

Molecular analysis of the recognition of human immunodeficiency virus type 1 (HIV-1) proteins by human B and T lymphocytes is essential for the development of immunotherapeutic strategies and vaccines against HIV-1 infection and for understanding the pathophysiology of acquired immunodeficiency syndrome (AIDS). One critical aspect of this analysis is the delineation of functionally important B and T cell epitopes in HIV-1 proteins. For B lymphocytes, epitopes can be either linear or conformational and, in the case of the HIV-1 envelope glycoprotein gp120, may be associated with neutralization of viral infectivity by the relevant antibodies. As is discussed by Laal and Zolla-Pazner [this volume] and in several recent reviews [1, 2], substantial progress has been made in the identification of epitopes recognized by neutralizing antibodies. In the case of T cells, epitopes are invariably linear regions of amino acid sequence that are presented to T cells in the peptide-binding sites of major histocompatibility complex (MHC) molecules. Rapid progress has been made in the identification of epitopes recognized by CD8+ cytolytic T lymphocytes (CTLs), which contribute to antiviral responses by lysing infected host cells. Studies of CD8+ T cell epitopes in HIV-1 proteins are discussed by Phillips and McMichael [this volume] and in recent reviews [3, 4]. The present chapter deals with HIV-1 epitopes recognized by the other major subset of T cells, the CD4+ T cells.

General Aspects of the CD4+ T Cell Response to HIV-1

By producing cytokines that activate B lymphocytes, macrophages, CD8+ T cells and other cell types, CD4+ T cells play a central role in most immune responses. Some CD4+ T cells also have cytolytic activity and can lyse virally infected cells (see below). In contrast to CD8+ T cells which recognize viral antigens synthesized in infected cells, CD4+ T cells typically recognize antigens that are derived from the extracellular environment. Recognition requires uptake by antigen-presenting cells (APCs) such as macrophages, dendritic cells and antigen-specific B lymphocytes. Following uptake, antigens are degraded in a low-pH endocytic compartment and then peptide fragments of the antigen are displayed on the surface of the APCs bound to class II MHC molecules for subsequent recognition by CD4+ T cells. Understanding the CD4+ T cell response to HIV-1 at the molecular level requires determining which peptide fragments of HIV-1 proteins are presented.

The Importance of Defining CD4+ T Cell Epitopes

The delineation of CD4+ T cell epitopes in HIV-1 proteins is important for several aspects of AIDS research. The development of vaccines that elicit strong neutralizing antibody responses is expected to require that the vaccine elicit a good CD4+ helper T cell response as well. Understanding the regions of the envelope protein that are recognized by CD4+ T cells may thus facilitate vaccine design. A major problem in the development of HIV-1 vaccines is the enormous degree of interisolate sequence variability [5–7]. As is discussed below, identification of T cell epitopes in HIV-1 proteins permits a molecular analysis of the effects of HIV-1 sequence variability on T cell recognition of these proteins. Obviously, for purposes of vaccine design, there is a need to identify T cell epitopes in highly conserved regions of HIV-1 proteins.

Analysis of CD4+ T cell epitopes in HIV-1 proteins is also important for determining whether disease progression is related to the continual generation in an infected individual of viral escape mutants that are not recognized by the HIV-1-specific CD4+ T cells present. Considerable evidence now indicates that multiple distinct but closely related viral clones are present in each infected individual and that these variants coevolve over time [8, 9]. The relatively high proportion of nonconservative nucleotide changes in the envelope protein argues strongly for immunological selection as a major drive force for genetic change. For the envelope protein, escape from neutralizing antibodies provides a basis for selection. For all HIV-1 proteins, escape from recognition by CD8+ CTLs may be a selective force. Recent work by Phillips and McMichael [this volume] suggests that *gag* mutations occurring in vivo may permit escape from *gag*-specific CD8+ CTLs. It is presently unclear to what extent CD4+ T cell responses provide selective pressure for the generation of escape mutants. Interestingly, many of the *env*-specific

CD4+ T cells cloned from seronegative vaccine recipients are cytolytic [10]. Whether HIV-1-specific CD4+ CTLs are part of the anti-HIV-1 response in natural infection is as yet unclear, but if so, then it is possible that mutations in CD4+ T cell epitopes may give rise to viral variants that escape this form of immune surveillance. Understanding of this problem may thus be increased by delineation of CD4+ T cell epitopes, particularly the longitudinal analysis of the CD4+ T cell response in a given individual to the viral strain infecting that individual.

Analysis of CD4+ T cell epitopes may also contribute to understanding of the interaction between HIV-1-specific CD4+ and infected cells. One of the unique features of HIV-1 infection is that cells that are productively infected in vivo all express class II MHC gene products. While it was originally proposed that only antigens taken up from the extracellular environment could be processed for association with class II MHC molecules [11], it has become increasingly clear that some viral proteins synthesized in infected cells can also be processed for association with class II MHC gene products. For example, recent studies indicate that the HIV-1 envelope glycoprotein, when synthesized in infected cells, is processed for association with class II MHC molecules [10, 12]. This processing reaction occurs in a post-endoplasmic reticulum compartment and may involve expression of the protein on the cell surface followed by internalization and processing by the normal class II pathway [K.C. Callahan, J.F. Rowell and R.F. Siliciano, unpubl. results]. Cells expressing endogenously synthesized forms of the HIV-1 *gag* proteins are also recognized by *gag*-specific CD4+ T cells [13], probably by a different mechanism. In any event, it is clear that CD4+ T cells can recognize processed HIV-1 proteins on the surface of infected cells, and thus the epitopes involved in this reaction should be identified.

The CD4+ T Cell Response to HIV-1 in Natural Infection

Because CD4+ T cells play a central role in most immune responses, it is likely that the CD4+ T cell response to HIV-1 contributes in an important way to controlling HIV-1 infection, particularly in the early stages of the infection before CD4+ T cell depletion has become severe and before qualitative defects in the CD4+ T cell response develop. Unfortunately, limited information is currently available regarding the CD4+ T cell response to HIV-1 proteins in infected individuals. In contrast to the CD8+ T cell response to HIV-1, which is readily measured in most HIV-1-infected individuals, the CD4+ T cell response is much more difficult to detect. Of course, there is a general impairment of CD4+ T cell responses in infected individuals. Some reports suggest a selective defect in the CD4+ T cell responses to HIV-1 antigens [14–16], but as CD4+ T cell depletion proceeds, CD4+ T cell responses to a variety of antigens become more difficult to detect [17]. This is due not only to the progressive decline in CD4+ T cell numbers but also to a qualitative defect in the functional activity of surviving CD4+ T cells. As was initially demonstrated by Lane et al. [18], this functional deficit is particularly prominent in AIDS patients whose surviving CD4+ T cells do not show proliferative responses to soluble protein recall antigens. Several mechanisms have been proposed for the functional defects in CD4+ T cells in infected individuals [reviewed in ref. 19 and 20].

Despite the impairment of CD4+ T cell responses that occurs as a result of HIV-1 infection, several laboratories have documented HIV-1-specific CD4+ T cell responses in a fraction of infected individuals. Peripheral-blood mononuclear cells (PBMCs) from some infected individuals show proliferative responses, presumably mediated by CD4+ T cells, to HIV-1 antigens including HIV-1 virions [15], purified HIV-1 proteins [14, 15, 21, 22] and synthetic peptides [23, 24]. *env*-specific [25] and *gag*-specific [13] CD4+ T cell clones have been isolated from seropositive individuals. In other studies, specialized techniques have been used to detect virus-specific CD4+ T cell responses in the context of natural infection. For example, proliferative responses to inactivated simian immunodeficiency virus (SIV) can be demonstrated using peripheral-blood lymphocytes from SIV-infected rhesus macaques and sooty mangabeys if antigen-presenting monocytes are first pulsed with inactivated virus and washed before the addition of responding T cells [26]. This approach overcomes the potential nonspecific inhibitory effects of inactivated virus or of viral proteins on CD4+ T cell responses. Antigen-induced interleukin-2 secretion has also been used to detect *env*-specific T cell responses in seropositive individuals whose proliferative responses to antigen are no longer detectable [27].

Analysis of CD4+ T cell response to HIV-1 proteins in seropositive individuals is complicated by HIV-1 sequence variability, which is particularly prominent within the gp120 coding sequence. Most studies of CD4+ T cell responses to HIV-1 have been carried out using as a stimulating antigen either synthetic peptides or recombinant proteins based on standard strains of HIV-1 such as LAI[1]. However, many human CD4+ T cell clones specific for gp120 from a given HIV-1 isolate cross-react poorly, if at all, with gp120 from unrelated isolates [10, 25, 29–31]. Therefore, gp120-specific CD4+ T cell responses will not be readily detected using stimulating antigen preparations from standard HIV-1 strains because of sequence differences between the isolate infecting a given individual and the isolate used in the preparation of the stimulating antigen. Even epitopes in relatively conserved regions of gp120 such as the putative CD4+-binding site [30] have a sufficient degree of variability to interfere with T cell recognition. Because all of the available information on CD4+ T cell responses to HIV-1 comes from studies in which proteins or peptides derived from standard strains of HIV-1 have been used as antigens, the real nature of the CD4+ T cell response to HIV-1 gp120 during natural infection is unknown.

For the reasons cited above, the CD4+ T cell response to HIV-1 in seropositive individuals has proven difficult to study at the molecular level. Nevertheless, substantial progress has been made in the analysis of CD4+ T cell epitopes in HIV-1 proteins. This work is reviewed below along with some very promising new methods for identifying the naturally processed epitopes in HIV-1 proteins.

Role of the Presenting MHC Molecule

The delineation of a T cell epitope at the biochemical level requires identification of the presenting MHC molecule as well as identification of the residues from the relevant protein antigen that are recognized by T cells in the context of that MHC molecule. Because polymorphic amino acids lining the peptide-binding site of a given MHC molecule determine which peptides can bind in this site [32, 33], a peptide that represents a T cell epitope with respect to a given MHC molecule will generally not be presented by other allelic variants of that MHC molecule. *In this sense, T cell epitopes have meaning only in the context of a given MHC molecule. Individual MHC alleles are typically expressed only by a small fraction of the population. Thus, while it is clearly important to identify T cell epitopes in HIV-1 proteins, such information generally applies only to a limited segment of the population, those who express the relevant MHC molecule.*

The situation is further complicated by the fact that there is a greater

[1] Throughout this article, HIV-1 isolate nomenclature and protein sequence numbering for HIV-1 isolates are based on the compilation by Meyers et al. [28]. As a result of analysis of the origin of early HIV-1 isolates (see table 1 for references), it now appears that the LAV-1 and IIIB isolates of HIV-1 both came from an isolate from patient LAI. Sequences originally designated as LAV-1 (BRU) are now designated LAI. Individual IIIB-related sequences retain their original specific designations (for example IIIB BH10 is herein referred to as LAI BH10).

degree of sequence polymorphism among human MHC molecules than is revealed by conventional HLA typing methods. This fact, together with the fact that even very subtle sequence differences in MHC molecules can have a dramatic effect on T cell recognition, must be considered before generalizations about a given epitope can be made. For example, the human DR4 serological specificity consists of the closely related class II molecules Dw4, Dw10, Dw13, Dw14 and Dw15, which differ from one another by one to five amino acids in the first domain of the β-chain [34, 35]. The human T cell clone Een217 [29], which is specific for HIV-1 gp120 amino acids 410–429 in the context of HLA-DR4, actually recognizes this epitope in the context of a relatively rare subtype of DR4, Dw10, carried by the donor from whom the clone was obtained [36]. Dw13 APCs can also present this peptide to clone Een217, but Dw4, Dw14 and Dw15 APCs cannot. In principle, the failure of the closely related MHC molecules to present this antigenic peptide may be the result of failure to bind the peptide. Alternatively, if binding does occur, the resulting peptide-MHC complexes may not be recognized by the T cell receptor on Een217 cells. In the former case, one would conclude that the gp120 peptide 410–429 is a T cell epitope with respect to Dw10, but not Dw4, Dw14 or Dw15. Interestingly, it appears that the 410–429 peptide can bind to these MHC molecules [36], suggesting that the problem is at the level of interactions with the T cell receptor on the Een217 cells. In this case, peptide 410–429 may or may not be a T cell epitope with respect to Dw4, Dw14 and Dw15, depending on whether or not T cells capable of recognizing the peptide in the context of these MHC molecules exist in individuals carrying those alleles. In either case, it would be inappropriate to conclude that peptide 410–429 is a DR4-restricted epitope based solely on the analysis of a single T cell clone. Analogous results have been obtained in a recent analysis of a DPw4.2-restricted human T cell clone specific for an epitope in gp41 [31]. The relevant peptide is presented by APCs expressing DPw4.2 but not by APCs expressing the closely related DP β-chain allele types DPw4.1 or DPw2.1, which differ from DPw4.2 by three and one amino acids, respectively [31]. These findings demonstrate that great caution must be used in making generalizations about T cell epitopes recognized in the context of MHC molecules for which multiple subtypes exist.

Although analysis of epitopes recognized by individual T cell clones gives information that is applicable only to the small percentage of the population carrying the relevant MHC allele, there are two situations in which information about T cell epitopes may be more generally relevant.

Some MHC alleles are expressed in a relatively high proportion of the population. Typically, such alleles are from loci for which there is a relatively limited degree of polymorphism, such as HLA-DQ or -DP. Identification of T cell epitopes recognized in the context of such alleles is important because the results are relevant to a larger fraction of the population. For example, the gp41 peptide 584–595 is recognized in the context of HLA-DPw4.2 which has a high allele frequency among Caucasians (27.2%). (It is important to remember that there are some striking racial differences in the frequency of particular MHC alleles.)

The other case involves T cell epitopes that are recognized in the context of multiple different MHC molecules [37, 38]. For example, Sinagaglia et al. [37] have shown that a peptide from the *Plasmodium falciparum* circumsporozoite protein can be presented by several different HLA-DR molecules to T cell clones from naive donors. For each clone, recognition showed absolute MHC restriction. The existence of these 'universal epitopes' is a surprising finding because, as is discussed above, polymorphic residues in MHC molecules can strongly influence the ability of a given peptide to associate with a given MHC molecule. It is possible that the circumsporozoite peptide associates mainly with the nonpolymorphic α-chain of these DR molecules. With respect to epitopes in HIV-1 proteins, Berzofsky et al. [39] have analyzed three T cell epitopes in the HIV-1 envelope protein and shown that these epitopes are recognized by CD8+ CTLs in association with three or four of ten MHC haplotypes tested. In this study, synthetic peptides were used in in vitro stimulations to elicit the peptide-reactive CTL populations, raising the possibility that rare clones capable of recognizing the peptide in the context of each MHC molecule were expanded. Such clones may not be prominent in natural infection. Nevertheless, they were shown to be capable of lysing target cells expressing endogenous envelope protein. It remains to be determined whether there are human CD4+ T cell epitopes that can be recognized in association with multiple class II molecules and, if so, whether they will be useful in vaccine development.

Approaches to Identifying CD4+ T Cell Epitopes

Several methods have been used to identify T cell epitopes in HIV-1 proteins. As discussed above, rigorous delineation of a T cell epitope requires identification not only of the antigenic peptide recognized but also identification of precisely which MHC molecule is involved in the presentation of that peptide. This can be best done with antigen-specific T cell clones. However, other approaches have also been used. The approaches that have been used to date and the epitopes identified using these methods are summarized below.

Methods for Predicting T Cell Epitopes

Several methods for predicting T cell epitopes have been proposed, including algorithms based on amphipathic α-helical propensity [40, 41] and general sequence patterns [42, 43]. Some potential epitopes in HIV-1 proteins identified by these methods have been shown to be recognized by HIV-1-specific T cells [41]. However, the reliability of such methods for predicting human T cell epitopes is not yet well established given that epitopes exist throughout the coding sequence of each HIV-1 protein (see below).

The most significant recent advance in the prediction of T cell epitopes has come from structural analysis of MHC molecules and of peptides eluted from MHC molecules. Recent crystallographic studies of class I MHC molecules have provided some important insights into the way in which antigenic peptides are presented to T cells by MHC molecules. For example, electron density present in the peptide-binding site of HLA-B27 appears to represent self-peptides in the form of nonamers bound in an extended, kinked β-strand [44]. Side chains from amino acid residues P2, P3, P7 and P9 of the peptide appear to bind to pockets in the floor of the B27 peptide-binding site, and thus the sequences of peptides that can be accommodated by this site are expected to show limitations at these positions. Structures of class II MHC molecules have not yet been solved at high resolution, but analysis of conserved sequence motifs suggests that the α1 and β1 domains of class II molecules may fold to form the same type of peptide-binding site that is formed by the N-terminal two domains of the class I heavy chain [45]. As is discussed below, the most obvious difference between class-I- and class-II-restricted T cell epitopes is in the length of the bound peptide. Peptides bound to class II molecules appear to be longer than the 8- or 9-residue peptides presented by class I molecules. In any event, it may ultimately be possible to usc structural information about individual MHC molecules to develop general rules about the sequences of peptides that can bind to particular MHC molecules. Crystallographic studies also provide a solid structural basis for analyzing notions about the nature of T cell epitopes. For example, the HLA-B27 data are inconsistent with the notion that T cell epitopes are amphipathic α-helices.

Future approaches to the delineation of T cell epitopes are likely to be based on the notion that the unique molecular architecture of each MHC molecule determines the possible range of peptides that it can present. Analysis of known epitopes recognized in the context of a particular MHC molecule has been used to derive MHC-allele-specific consensus sequences [46, 47]. As is discussed below, analysis of sequences of naturally processed peptides eluted from particular MHC molecules may eventually provide the best information about the sequence requirements for binding to particular MHC molecules. However, these sequence requirements are likely to be so broad that they may have limited predictive value except to exclude peptides lacking appropriate residues at critical anchor sites. At the present time, direct experimental delineation of T cell epitopes is the only reliable method for identifying them.

Clonal Analysis of Human CD4+ T Cell Epitopes

Studies with HIV-1-specific human T cell clones have provided the most detailed information about T cell epitopes in HIV-1 proteins. In general, these studies have utilized the classical approach for defining T cell epitopes, which involves the isolation of antigen-specific T cell clones, the identification of the appropriate restriction element using panels of APCs with different MHC genotypes and the localization of the epitope using proteolytic or recombinant fragments of the relevant antigen. Fine mapping of the epitope is then done with appropriate synthetic peptides. CD4+ human T cell clones specific for the HIV-1 envelope protein have been isolated from naive seronegative donors following in vitro immunization with recombinant envelope protein [29, 30, 48]. HIV-1-specific CD4+ T cell clones have also been isolated from AIDS vaccine recipients [10, 31, 49] and seropositive individuals [13, 25]. Epitopes that have been defined thus far using such clones are listed in table 1. As discussed above, a given antigenic peptide derived from an HIV-1 protein is a T cell epitope only in the context of the MHC molecule that presents this peptide. Thus, the complete delineation of a T cell epitope should also include identification of the MHC molecule involved in presenting that peptide. Table 1 includes only those epitopes for which the relevant MHC restriction elements have been determined.

The first human CD4+ T cell epitope to be fully delineated was an epitope in gp120 that is recognized by T cell clones isolated from a seronegative donor following in vitro immunization with recombinant LAI (PV22) gp120. This epitope (*env* amino acids 410–429) is recognized in the context of HLA-DR4 Dw10 [29, 30]. Interestingly, this epitope includes residues that have been implicated binding to CD4 [53]. Although there is a conserved

Table 1. CD4+ T cell epitopes in HIV-1 proteins

HIV-1 protein	Isolate[1]	Epitope	Sequence	MHC restriction	Comment	Reference
gag	LAI	140–148	GQMVHQAIS	DQwl and 3	Conserved epitope	13
env	SF2	292–300	NESVAINCT[2]	DR2 (w15)	Glycosylation prevents recognition	50
env	LAI (PV22)	410–429	GSDTITLPCRI-KQFINMWQE	DR4(Dw10)	Overlaps CD4-binding site	30
env	LAI	584–595	RILAVERYLKDQ	DPw4.2	Conserved; also a B cell and CD8+ CTL epitope	31

[1] Isolate nomenclature and sequence numbering are from the compilation by Meyers et al. [28]. As a result of analysis of the origin of early HIV-1 isolates [51, 52], it now appears that the LAV-1 and IIIB isolates of HIV-1 both came from an isolate from patient LAI. Sequences originally designated as LAV-1 (BRU) are now designated LAI. Individual IIIB-related sequences retain their original specific designations (for example IIIB PV22 and IIIB BH10 are herein referred to as LAI PV22 and LAI BH10, respectively).
[2] Asn residues in bold are N-linked glycosylation sites.

core of residues within this epitope, substantial interisolate sequence variability is observed. Callahan et al. [30] used synthetic peptides representing the relevant portion of gp120 from various isolates of HIV-1 to examine how this sequence variation affected recognition by clones generated in response to LAI gp120. Proliferative responses of DR4-Dw10-restricted clones to these peptides were compared to responses to the corresponding peptide from the isolate used for immunization (table 2). In general, variant peptides were recognized poorly compared to the reference LAI peptide. Peptide competition experiments were used to show that variation affected recognition at two levels. For some strains, variation in the DR4 Dw10 epitope was sufficient to alter the interaction of antigen receptors on gp120-specific human T cell clones with peptide-DR4 complexes on APCs. In the case of two Zairean isolate strains, the natural variation was sufficient to prevent the critical initial interaction between the relevant gp120 peptide and DR4 Dw10 on the APCs. However, these strains were highly divergent from the reference strain. Thus, it is encouraging to note that the range of natural sequence variation in this T cell epitope falls for the most part within the

Table 2. Effect of HIV-1 sequence variability on recognition by the DR4 (Dw10)-restricted clone Een217 specific for LAI (PV22) gp120 (adapted from Callahan et al. [30])

Isolate/ clone	Origin[1]	Sequence[2] (410–429)	Sequence distance from LAI[3]	Recognition by Een217[4]	Binding to DR4[5]
PV22	NA	GS DTI TLPCRI KQFI NMWQ E	–[6]	+++	+
BH10	NA	-------------I------	–	++	+
HXB2	NA	-------------I-----K	–	+++	+
SF2	NA	-N---I-------I------	236	+	+
RF	NA	-N-----------IV-----	323	++	+
WMJ2	NA	N-TL---------I-----G	274	+	+
NY5	NA	NNE--II------I------	266	++	+
MAL	Z	STGS---------I-----K	412	+	+
CDC42	NA	TG-I---------I--R--V	241	–	+
Z3	Z	TGN---------VVRT--G	504	–	–
ELI	Z	TNTN---Q-----I-K-VAG	467	–	–

[1] North American (NA) or Zairean (Z) isolates of HIV-1.
[2] Amino acid sequence of the envelope protein from the indicated isolate in the region of the DR4-Dw10-restricted T cell epitope (residues 410–429 of the PV22 gp120 sequence). This table includes a set of highly related HIV-1 LAI clones (PV22, BH10 and HXB2) derived from a single infected cell line, the H9-HTLV-IIIB line. In addition, more divergent HIV-1 strains obtained from different infected individuals from diverse geographic areas are included. For a compilation of sequence data and references, see Meyers et al. [28].
[3] As a first approximation of the phylogenetic relationship between HIV-1 isolates, computer-generated phylogenetic trees produced by Meyers et al. [28] have been used to derive minimal sequence distances in units of nucleotide substitutions between the indicated strains. This analysis, which is based on the nucleotide sequence of the entire *env* gene, provides an indication of the degree of relatedness of various HIV-1 isolates.
[4] Summary of results of proliferation assays using the T cell clone Een217 which is specific for the PV22 epitope 410–429 and DR4 Dw10 APCs. Results are expressed in terms of the concentration of peptide required during pulsing of DR4-Dw10-expressing APCs to give 50% maximal proliferation of Een217: +++ = <0.1 μ*M*; ++ = 0.1–1 μ*M*; + = 1–3 μ*M*; – = >30 μ*M*.
[5] Binding to DR4 was considered positive if the peptide either induced a DR4-restricted T cell proliferative response or inhibited responses to the PV22 peptide.
[6] Sequence distances are not calculated for the closely related, IIIB-derived molecular clones.

range of peptide sequences that can be accommodated by DR4 Dw10. Nevertheless, T cell clones specific for the LAI sequence showed poor cross-reactivity with most other strains.

With regard to pathogenesis, these data suggest that the low level of variability occurring within a given infected individual will generally *not* be sufficient to abolish the interaction of appropriate T cell epitopes with MHC molecules. However, even very minor sequence changes in a T cell epitope can lead to diminished recognition at the level of the T cell receptor. This may enable variant viral clones to evade ongoing T cell responses. Nevertheless, as long as the epitope can associate with appropriate MHC molecules expressed in that individual, de novo responses to the variant can in principle be generated provided that the relevant T cell compartment remains functionally intact.

These results are also relevant to vaccine development. All but the most highly divergent HIV-1 strains contain a gp120 epitope capable of interacting with DR4 Dw10. Individuals carrying this allele who are immunized with recombinant gp120 from a given viral strain such as LAI may mount a CD4+, gp120-specific secondary T cell response following infection with certain other HIV-1 strains that have few sequence changes in this gp120 epitope. For some responding clones, the magnitude of the response is likely to be substantially reduced due to minor sequence changes which decrease the affinity of T cell receptor binding. However, immunization may induce a number of different clones which recognize the same epitope in a DR4-restricted fashion but which have slightly different antigen fine specificities. Some of these clones may respond very well to particular variants. In the case of more highly divergent strains, no secondary response directed at the epitope will be elicited because residues critical for interaction with T cell receptors on most LAI-induced clones may be absent. Therefore, with respect to this epitope, little protective effect may be expected. In the case of infection with highly divergent isolates unable to bind to DR4, there will be no DR4-restricted response directed at this epitope, and responsiveness to gp120 will depend on presence or absence in the host of other restriction elements capable of interacting with some processed peptide from the infecting strain.

The studies described above highlight the difficulties posed by HIV-1 sequence variability. With respect to vaccine development, it is essential to identify epitopes in conserved regions of HIV-1 proteins. A human CD4+ CTL epitope located in a region of the HIV-1 envelope protein gp41 (584–595) that is highly conserved among various HIV-1 strains has recently been described [31]. This epitope is recognized by CD4+ CTL clones that were induced in seronegative humans by immunization with recombinant gp160. Fusion proteins carrying portions of the HIV-1 *env* gene and synthetic peptides were used to localize this epitope to amino acids 584–595 of the LAI *env* sequence (table 1). Only two positions within this epitope showed variation among North American HIV-1 isolates, and the substitutions were conservative in nature. The Ile to Val change at position 585 in this strain does not strongly affect recognition of this epitope by DPw4.2-restricted T cell clones. There is also a Lys to Arg substitution at position 593 in about half of the known HIV-1 isolates. This substitution abolished recognition, probably by interfering with the peptide-MHC interactions. Thus, vaccines containing this epitope may induce a CD4+ T cell response active against at least half of all HIV-1 strains in those carrying this allele. Interestingly, this epitope is recognized in association with one subtype of the widely distributed human class II MHC specificity DPw4, namely DPw4.2. The relatively high frequency of this allele (27.2% among Caucasians) makes it likely that a larger fraction of the population would generate a response directed at this epitope than would be the case for epitopes recognized in the context of gene products of most other class II and class I loci. As discussed above, the closely related DP β-chain allele types 4.1 and 2.1 were unable to present this gp41 peptide to DPw4.2-restricted clones. Comparison of the structure of this epitope with that of other peptides recognized in the context of DPw4.2 led to the identification of a consensus sequence for DPw4.2-binding peptides. Because this gp41 CTL epitope is highly conserved and is recognized in the context of a common DP allele, it may represent an important target region for vaccine development. This region of gp41 is also a major B cell epitope and is recognized by CD8+ CTL [see chapters by A.R. Neurath and F. Chiodi et al., this volume].

An interesting problem in the analysis of T cell epitopes in the envelope glycoprotein is the effect of N-linked glycosylation on T cell recognition. There are 24–26 N-linked glycosylation sites in gp120. Botarelli et al. [50] have recently shown that a subset of human CD4+ T cell clones isolated from seronegative volunteers immunized with a nonglycosylated form of gp120 fail to respond to glycosylated forms of the protein. The epitope recognized by one such clone (**N**E**S**VAI**N**C**T**) contains two N-linked glycosylation sites (indicated in bold). Synthetic peptides and gp120 expressed in a nonglycosylated form are recognized by the clone in the context of HLA-DR2(w15), while native gp120 is not. The data of Botarelli et al. are consistent with the notion that glycosylated versions of the peptide either do not bind to DR2, or if bound, do not interact with T cell receptors on clones specific for this epitope in nonglycosylated gp120. The data also suggest that the enzymes that hydrolyze glycoproteins in the cellular compartment where processing of exogenous antigens occurs do not do so in a way that generates peptides equivalent to those generated during processing of the nonglycosylated forms of the protein. These data are relevant to the development of synthetic peptide or nonglycosylated recombinant protein vaccines based on the HIV-1 envelope protein because they indicate that some of the CD4+ T cell clones induced by such vaccines may not recognize cells presenting viral protein.

Analysis of Peptide-Specific T Cell Responses in Infected Individuals

Although the analysis of HIV-1-specific T cell clones can provide definitive information about T cell recognition of particular epitopes, it does not give a complete picture of the antigenicity of HIV-1 proteins. Clonal analysis is subject to sampling errors such as the failure to detect epitopes for which the relevant clones have not been isolated. An alternative approach to the analysis of T cell epitopes involves analysis of the responses of polyclonal T cell populations from infected individuals to various HIV-1 antigen preparations. In particular, synthetic peptides have been used to stimulate proliferative responses and cytokine production by uncloned PBMCs from HIV-1-seropositive donors [23, 24, 27, 54, 55]. This approach has the advantage that it detects epitopes actually recognized by responding T cells activated during natural infection. It is important to note that T cell responses to HIV-1 peptides are generally not detected using PBMCs from normal HIV-1-seronegative donors.

Probable CD4+ T cell epitopes identified in this manner are listed in table 3. One caveat is that in these studies the phenotype of the responding cells is often not determined. Another problem with this approach is that without cloning it is difficult to ascertain the MHC restriction elements involved in presenting a given peptide. A final problem is that in many cases the magnitude of the response is quite low. For example, in most of these studies, a stimulation index of 2 is used as a cutoff for a positive response. The significance of weak responses of this kind is unclear.

Despite these problems, this approach has provided a number of new insights into the nature of T cell epitopes in HIV-1 proteins. The approach has been particularly useful in demonstrating that T cell epitopes exist throughout the entire coding sequence of each HIV-1 protein examined. The epitopes recognized in a given individual depend of course on the MHC genotype of that individual. For example, in a study of 21 synthetic peptides based on LAI *gag, pol* and *env* sequences, Schrier et al. [23] found that all but one of the peptides stimulated a proliferative response in at least one of the 29 seropositive individuals studied. These peptides were selected on the basis of hydrophilicity and antibody reactivity, and yet all but one served as a

Table 3. Peptides that elicit T cell responses in HIV-1-seropositive individuals

HIV-1 protein	Isolate[1]	Epitope	Sequence	Type of response[2]	Frequency of response[3]	Reference
gag (p17)	LAI	22–29	RPGGKKKY	P	14% (4/29)	23
	LAI (BH10)	33–47	HIVWASRELERFAVN	P	56% (9/16)	24
	LAI (BH10)	93–107	EIKDTKEALDKIEEE	P	25% (4/16)	24
	LAI (BH10)	118–132	AAADTGHSSQVSQNY	P	25% (4/16)	24
gag (p24)	LAI (BH10)	133–147	PIVQNIQGOMVHQAI	P	63% (10/16)	24
	LAI (BH10)	208–222	EAAEWDRVHPVHAGP	P	38% (6/16)	24
	LAI	228–235	MREPRGSD	P	14% (4/29)	23
	LAI (BH10)	228–242	MREPRGSDIAGTTST	P	38% (6/16)	24
	LAI (BH10)	278–292	SPTSILDIRQGPKEP	P	50% (8/16)	24
	LAI	282–301	ILDIRQGPKEPFRDY-VDRFY	P	24% (7/29)	23
	LAI (BH10)	288–302	GPKEPFRDYVDRFYK	P	50% (8/16)	24
gag (p15)	LAI (BH10)	393–407	FNCGKEGHTARNCRA	P	31% (5/16)	24
	LAI (BH10)	418–432	KEGHQMKDCTERQAN	P	31% (5/16)	24
	LAI (BH10)	423–437	MKDCTERQANFLGKI	P	38% (6/16)	24
	LAI	439–446	PSYKGRPG	P	14% (4/29)	23
gag (p6)	LAI	478–485	GVETTTPS	P	14% (4/29)	23
	LAI (BH10)	473–487	ESFRSGVETTTPPQK	P	50% (8/16)	24
pol (RT)	LAI	205–219	CTEMEKEGKISKIGP	I	53% (9/17)	54
pol (p32)	LAI	732–742	IDKAQEEHEKY	P	34% (10/29)	23
	LAI	911–925	NFKRKGGIGGYSAGE	P	28% (8/29)	23
	LAI	935–949	IQTKELQKQITKIQN	P	38% (11/29)	23
	LAI	954–966	YRDNKDPLWKGPA	P	48% (14/29)	23
env (gp120)	LAI	75–85	VPTDPNPQEVV	P	7% (3/29)	23
	LAI	108–118	IISLWDQSLKP	P	31% (9/29)	23
	LAI (BH10)	112–124	HEDIISLWDQSLK	P	11% (4/35)	27
				I	29% (12/42)	
	LAI	115–125	SLKPCVKLTPL	P	38% (11/29)	23
	LAI	234–244	NNKTFNGKGPC	P	7% (3/29)	23
	LAI (BH10)	269–283	EVVIRSANFTDNAKT	P	30% (6/20)	24
	LAI (BH10)	274–288	SANFTDNAKTIIVQL	P	40% (8/20)	24
	LAI	302–312	TRPNNNTRKSI	P	14% (4/29)	23
	LAI (BH10)	308–322	RIQRGPGRAFVTIGK	P	5% (1/22)	27
				I	21% (6/29)	
	LAI (BH10)	309–323	IQRGPGRAFVTIGKI	P	30% (6/20)	24
	LAI (BH10)	314–328	GRAFVTIGKIGNMRQ	P	35% (7/20)	24
	SF2	346–359	QIVKKLREQFGNNK	P	10% (1/10)[4]	55
	LAI (BH10)	364–378	SSGGDPEIVTHSFNC	P	40% (8/20)	24
	LAI	368–377	QSSGGDPEIV	P	21% (6/29)	23

Table 3. (cont.)

HIV-1 protein	Isolate[1]	Epitope	Sequence	Type of response[2]	Frequency of response[3]	Reference
env (gp120) (cont.)	LAI (BH10)	369–383	PEIVTHSFNCGGEFF	P	40% (8/20)	24
	LAI (BH10)	394–408	TWFNSTWSTKGSNNT	P	55% (11/20)	24
	LAI (BH10)	399–413	TWSTKGSNNTEGSDT	P	35% (7/20)	24
	LAI (BH10)	421–436	KQIINMWQEVGKAM-YA	P	11% (4/35)	27
				I	33% (14/42)	
	LAI (BH10)	424–438	INMWQEVGKAMYAPP	P	30% (6/20)	24
	LAI (BH10)	459–473	GNSNNESEIFRPGGG	P	45% (9/20)	24
	SF2	466–481	FRPGGGDMRDNWRS-EL	P	15% (2/13)[4]	55
	LAI (BH10)	474–488	DMRDNWRSELYKYKV	P	40% (8/20)	24
	LAI	474–486	RPGGGDMRDNWRS	P	0% (0/29)	23
	LAI (BH10)	484–498	YKYKVVKIEPLGVAP	P	40% (8/20)	24
env (gp41)	LAI	584–609	RILAVERYLKDQQLL-GIWGCSGKLIC	P	24% (7/29)	23
	LAI (BH10)	547–561	GIVQQQNNLLRAIEA	P	13% (3/23)	24
	LAI (BH10)	562–576	QQHLLQLTVWGIKQL	P	39% (9/23)	24
	LAI	604–614	SGKLICTTAVP	P	41% (12/29)	23
	LAI	610–620	TTAVPWNAASWS	P	48% (14/29)	23
	LAI (BH10)	647–661	EESQNQQEKNEQELL	P	26% (6/23)	24
	LAI	655–667	QNQQEKNEQELLE	P	38% (11/29)	23
	LAI (BH10)	667–681	ASLWNWFNITNWLWY	P	48% (11/23)	24
	LAI (BH10)	682–696	IKLFIMIVGGLVGLR	P	26% (6/23)	24
	LAI	737–749	GIEEEGGERDRDR	P	21% (6/29)	23
	LAI (BH10)	827–841	DRVIEVQGAYRAIR	P	5% (1/22)	27
				I	21% (6/29)	
	LAI (BH10)	842–856	HIPRRIRQGLERILL	P	39% (9/23)	24

[1] Isolate nomenclature and sequence numbering are from the compilation by Meyers et al. [28].
[2] P = Proliferative; I = interleukin-2 production.
[3] Percent of seropositive individuals with a peptide-specific T cell response. Numbers in parentheses are the actual number of responders over the number of donors tested.
[4] A higher frequency of responses was observed when the peptides were conjugated to liposomes [55].

T cell epitope. Similar results were obtained by Wahren et al. [24] using a more extensive set of *gag* and *env* peptides. By measuring the proliferative response of PBMCs from infected individuals to these peptides, these authors identified epitopes unique to individual donors as well as epitopes that were recognized by a sizable fraction of the population tested. The peptides that most commonly elicited proliferative responses are included in table 3.

Taken together, the studies described above support the notion that potential T cell epitopes are located throughout the HIV-1 proteins examined to date and that the precise epitopes recognized by T cells from a given individual depend on the MHC genotype of that individual and possibly other factors affecting antigen processing. Support for these conclusions comes from an interesting recent study by Manca et al. [48]. These investigators found that peripheral-blood T cells from a seronegative volunteer, after in vitro stimulation with pools of gp120 peptides, responded to peptides spanning virtually the entire gp120 molecule. However, in vitro immunization with intact gp120 elicited a response that was focused on a particular epitope which varied from donor to donor [48].

Delineation of T Cell Epitopes in Animal Models

Several laboratories have identified T cell epitopes in HIV-1 proteins by immunizing experimental animals with the relevant proteins and then defining the epitopes recognized by the responding T cells [41, 56–58]. While it is certainly possible that epitopes identified in animal models will also be recognized by HIV-1-specific human T cells in the context of some MHC molecules, there is as yet no compelling evidence that animal models provide a reliable means for identifying epitopes that will be recognized in the context of particular human MHC molecules.

Delineation of Epitopes by Sequence Analysis of Peptides Eluted from MHC Molecules

Recent technical advances in the analysis of peptides bound to MHC molecules have given rise to an important new approach for the identification of T cell epitopes. This approach involves the treatment of whole cells or of affinity-purified MHC molecules with low pH to elute naturally processed peptides bound to MHC molecules. The eluted peptides are separated by HPLC and subjected to sequence analysis by Edman degradation [59–61], mass spectrometry [62] or by analytical HPLC using defined synthetic peptide standards [59, 63]. This approach has been successfully used in the identification of naturally processed self-peptides bound to class I [60] and class II [61] MHC molecules and in studies demonstrating the involvement of peptides in alloreactivity [64]. Recently, naturally processed viral peptides bound to class I MHC molecules expressed on infected cells have been identified by this approach [59, 63]. For two different murine class I molecules, the influenza virus peptides identified were nonamers [63], while a vesicular stomatitis virus peptide was found to be an octamer [59]. Comparison of the sequences of naturally processed influenza virus peptides with sequences of other peptides known to be presented in association with the relevant MHC molecules enabled the delineation of allele-specific consensus sequences [63]. Allele-specific motifs have also been derived by sequencing mixtures of self-peptides eluted from MHC molecules [65].

In principle, this approach could be applied to the identification of class-II-restricted T cell epitopes in HIV-1 proteins. One strategy would involve the exposure of appropriate class-II-expressing APCs to an HIV-1 protein, followed by acid elution of naturally processed peptides derived from the relevant protein. Another approach would involve the elution of peptides from HIV-1-infected cells. Previous studies have shown that endogenously synthesized forms of the envelope protein can be processed in infected cells for association with class II MHC molecules [10, 12]. The principal advantage of this approach is that it would identify the exact form of the antigen that is presented to T cells in vivo. The technique is sufficiently sensitive to detect peptides that are presented in only a few hundred copies per cell. However, the approach has some disadvantages. First, the demonstration that a given HIV-1 peptide can be eluted from a given class II molecule does not necessarily prove that the peptide is recognized by T cells, only that it is a potential epitope. Additional studies demonstrating T cell recognition of the peptide are required to prove that the relevant sequence is actually an epitope. Another problem is that large numbers of cells are required. Analysis by mass spectrometry can be done with 10^9 cells. Large numbers of cells are also needed for peptide identification by analytical HPLC, an approach which requires some prior knowledge about the epitope. Sequencing peptides by Edman degradation requires an even larger number of cells (10^{11}). Nevertheless, it is very likely that significant technical advances in the next few years will make this approach more feasible.

Conclusions

Despite the difficulties inherent in analyzing CD4+ T cell responses to HIV-1, some progress has been made in the identification of CD4+ T cell epitopes in HIV-1 proteins. Studies with responding cells from infected individuals indicate that T cell epitopes exist throughout the coding sequences of the HIV-1 proteins examined to date. Detailed studies with human T cell clones have documented the extraordinary specificity of the interactions of HIV-1 peptides, MHC molecules and T cell receptors. The most rigorous method for identifying epitopes involves the use of clones of known restriction specificity, but the results from epitope mapping studies of this kind generally apply only to individuals carrying the relevant MHC allele. Future approaches involving sequence analysis of naturally processed peptides may give the most definitive picture of the human T cell epitopes in HIV-1 proteins.

References

1 Goudsmit J, Back NKT, Nara PL: Genomic diversity and antigenic variation of HIV-1: Links between pathogenesis, epidemiology and vaccine development. FASEB J 1991;5:2427–2436.

2 Nara PL, Garrity RR, Goudsmit J: Neutralization of HIV-1: A paradox of humoral proportions. FASEB J 1991;5:2437–2455.

3 Mills KHG, Nixon DF, McMichael AJ: T-cell strategies in AIDS vaccines: MHC-restricted T-cell responses to HIV proteins. AIDS 1989;3:S101–S110.

4 Nixon DF, McMichael AJ: Cytotoxic T cell recognition of HIV proteins and peptides. AIDS 1991;5:1049–1059.

5 Wong-Staal F, Shaw GM, Hahn BH, Salahuddin SZ, Popovic M, Markham P, Redfield R, Gallo RC: Genomic diversity of human T-lymphotropic virus type III (HTLV-III). Science 1985;229:759–762.

6 Benn S, Rutledge R, Folks T, Gold J, Baker L, McCormick J, Feorino P, Piot P, Quinn T, Martin M: Genomic heterogeneity of AIDS retroviral isolates from North America and Zaire. Science 1985;230:949–951.

7 Hahn BH, Gonda AM, Shaw GM, Popovic M, Hoxie JA, Gallo RC, Wong-Staal F: Genomic diversity of the acquired immune deficiency syndrome virus HTLV-III: Different viruses exhibit greatest divergence in their envelope genes. Proc Natl Acad Sci USA 1985;82:4813–4817.

8 Hahn BH, Shaw GM, Taylor ME, Redfield RR, Markham PD, Salahuddin SZ, Wong-Staal F, Gallo RC, Parks ES, Parks WP: Genetic variation of HTLV-III/LAV over time in patients with AIDS or at risk for AIDS. Science 1986;232:1548–1553.

9 Saag MS, Hahn BH, Gibbons J, Li Y, Parks ES, Parks WP, Shaw GM: Extensive variation of human immunodeficiency virus type-1 in vivo. Nature 1988;334:440–444.

10 Orentas RJ, Hildreth JKE, Obah E, Polydefkis M, Smith GE, Clements ML, Siliciano RF: An HIV envelope protein vaccine induces CD4+ human cytolytic T cells active against HIV-infected cells. Science 1990;248:1234–1237.

11 Morrison LA, Lukacher AE, Braciale VL, Fan D, Braciale TJ: Differences in antigen presentation to MHC class I- and class II-restricted influenza virus-specific cytolytic T lymphocyte clones. J Exp Med 1986;163:903–917.

12 Polydefkis M, Koenig S, Flexner C, Obah E, Gebo K, Chakrabarti S, Earl PL, Moss B, Siliciano RF: Anchor-sequence dependent endogenous processing of the HIV-1 envelope glycoprotein gp160 for CD4+ T cell recognition. J Exp Med 1990;171:875–888.

13 Littaua RA, Oldstone MBA, Takeda A, Ennis FA: A CD4+ cytotoxic T-lymphocyte clone to a conserved epitope on human immunodeficiency virus type 1 p24: Cytotoxic activity and secretion of interleukin-2 and interleukin-6. J Virol 1992;66:608–611.

14 Krohn K, Robey WG, Putney S, Arthur L, Nara P, Fischinger P, Gallo RC, Wong-Staal F, Ranki A: Specific cellular immune response and neutralizing antibodies in goats immunized with native or recombinant envelope proteins derived from human T-lymphotropic virus type IIIB and in human immunodeficiency virus-infected men. Proc Natl Acad Sci USA 1987;84:4994–4998.

15 Wahren B, Morfeldt-Mansson L, Biberfeld G, Moberg L, Sonnerborg A, Ljungman P, Werner A, Kurth R, Gallo R, Bolognesi D: Characteristics of the specific cell-mediated immune response in human immunodeficiency virus infection. J Virol 1987;61:2017–2023.

16 Ranki A, Mattinen S, Yarchoan R, Broder S, Ghrayeb J, Lahdevirta J, Krohn K: T-cell response towards HIV in infected individuals with and without zidovudine therapy, and in HIV-exposed sexual partners. AIDS 1989;3:63–69.

17 Clerici M, Stocks NI, Zajac RA, Boswell RN, Lucey DR, Via CS, Shearer GM: Detection of three distinct patterns of T helper cell dysfunction in symptomatic, human immunodeficiency virus-seropositive patients. J Clin Invest 1989;84:1892–1899.

18 Lane HC, Depper JM, Greene WC, Whalen G, Waldmann TA, Fauci AS: Qualitative analysis of immune function in patients with the acquired immunodeficiency syndrome: Evidence for a selective defect in soluble antigen recognition. N Engl J Med 1985;313:79–84.

19 Rosenberg ZF, Fauci AS: The immunopathogenesis of HIV infection. Adv Immunol 1989;46:377–405.

20 Bollinger RC, Siliciano RF: Immunodeficiency in HIV infection; in Wormser GP (ed): AIDS and Other Manifestations of HIV Infection. New York, Raven Press, 1992, pp 145–163.

21 Torseth JW, Berman PW, Merigan TC: Recombinant HIV structural proteins detect specific cellular immunity in vitro in infected individuals. AIDS Res Hum Retroviruses 1988;4:23–30.

22 Borkowsky W, Krasinski K, Moore T, Papaevangelou V: Lymphocyte proliferative responses to HIV-1 envelope and core antigens by infected and uninfected adults and children. AIDS Res Hum Retroviruses 1990;6:673–678.

23 Schrier RD, Gnann JW Jr, Landes R, Lockshin C, Richman DJ, McCutchan A, Kennedy C, Oldstone MBA, Nelson JA: T cell recognition of HIV synthetic peptides in a natural infection. J Immunol 1988;142:1166–1176.

24 Wahren B, Rosen J, Sandstrom E, Mathiesen T, Modrow S, Wigzell H: HIV-1 peptides induce a proliferative response in lymphocytes from infected individuals. J AIDS 1989;4:448–456.

25 Walker CM, Steimer KS, Rosenthal KL, Levy JA: Identification of human immunodeficiency virus (HIV) envelope type-specific T helper cells in an HIV-infected individual. J Clin Invest 1988;82:2172–2175.

26 Ahmed-Ansar A, Powell JD, Jensen PE, Yehuda-Cohen T, McClure HM, Anderson D, Fultz PN, Sell KW: Requirements for simian immunodeficiency virus antigen-specific in vitro proliferation of T cells from infected rhesus macaques and sooty mangabeys. AIDS 1990;4:399–407.

27 Clerici M, Stocks NI, Zajac RA, Boswell RN, Berstein DC, Mann DL, Shearer GM, Berzofsky JA: Interleukin-2 production used to detect antigen peptide recognition by T-helper lymphocytes from asymptomatic HIV-seropositive individuals. Nature 1989;339:383–385.

28 Meyers G, Korber B, Berzofsky JA, Smith RF, Pavlakis GN: Human Retroviruses and AIDS. Los Alamos, Los Alamos National Laboratory, 1991.

29 Siliciano RF, Lawton T, Knall C, Karr RC, Berman P, Gregory T, Reinherz EL: Analysis of host-virus interactions in AIDS with anti-gp120 human T cell clones: Effect of HIV sequence variation and a mechanism for CD4+ cell depletion. Cell 1988;54:561–575.

30 Callahan KM, Fort M, Obah EA, Reinherz EL, Siliciano RF: Genetic variability in HIV-1 gp120 affects interactions with HLA molecules and T cell receptors. J Immunol 1990;144:3341–3346.

31 Hammond SA, Obah E, Stanhope PE, Montell CR, Strand M, Robbins FM, Bias WB, Karr RW, Koenig S, Siliciano RF: Characterization of a conserved epitope in HIV-1 gp41 recognized by vaccine-induced cytolytic T cells. J Immunol 1991;146:1470–1477.

32 Bjorkman PJ, Saper MA, Samraoui B, Bennett WS, Strominger JL, Wiley DC: Structure of the human class I histocompatibility antigen, HLA-A2. Nature 1987;329:506–512.

33 Bjorkman PJ, Saper MA, Samraoui B, Bennett WS, Strominger JL, Wiley DC: The foreign antigen binding site and T cell recognition regions of class I histocompatibility antigens. Nature 1987;329:512–518.

34 Cairns JS, Curtsinger JM, Dahl CA, Freeman S, Alter BJ, Bach FH: Sequence polymorphism of HLA DRβ1 alleles relating to T cell-recognized determinants. Nature 1985;317:166–168.

35 Gregersn PK, Shen M, Song Q-L, Merryman P, Degar S, Seki T, Maccari J, Goldbert D, Murphy H, Schwenzer J, Wang CY, Winchester RJ, Nepom GT, Silver J: Molecular diversity of HLA-DR4 haplotypes. Proc Natl Acad Sci USA 1986;83:2642–2646.

36 Olson RR, McNicholl J, Alber C, Klohe E, Callahan K, Siliciano RF, Karr RW: Acidic amino acid residues of HLA-DR4β determine T cell recognition of an HIV-1 gp120 peptide. Submitted.

37 Sinagaglia F, Guttinger M, Kilgus J, Doran DM, Matile H, Etlinger H, Trzeciak A, Gillessen D, Pink JRL: A malaria T-cell epitope recognized in association with most mouse and human MHC class II molecules. Nature 1988;336:788–790.

38 Panina-Bordignon P, Tan A, Termijtelen A, Demotz S, Corradin G, Lanzavecchia A: Universally immunogenic T cell epitopes: Promiscuous binding to human MHC class II molecules and promiscuous recognition by T cells. Eur J Immunol 1989;19:2237–2242.

39 Berzofsky JA, Cease KB, Cornette JL, Spouge JL, Margalit H, Berkower IJ, Good MF, Miller LH, DeLisi C: Protein antigenic structures recognized by T cells: Potential applications to vaccine design. Immunol Rev 1987;98:9–52.

40 Berkower I, Buckenmeyer GK, Berzofsky JA: Molecular mapping of a histocompatibility-restricted immunodominant T cell epitope with synthetic and natural peptides: Implications for T cell antigenic structure. J Immunol 1986;136:2498–2503.

41 Cease KB, Margalit H, Cornette JL, Putney SD, Robey WG, Ouyang C, Streicher HZ, Fischinger PJ, Gallo RC, DeLisi C, Berzofsky JA: Helper T-cell antigenic site identification in the acquired immunodeficiency syndrome virus gp120 envelope protein and induction of immunity in mice to the native protein using a 16-residue synthetic peptide. Proc Natl Acad Sci USA 1987;84:4249–4253.

42 Rothbard J: Peptides and the cellular immune response. Ann Int Past 1986; 137E:518–526.

43 Lamb JR, Ivanyi J, Rees ADM, Rothbard JB, Howland K, Young RA, Young DB: Mapping of T cell epitopes using recombinant antigens and synthetic peptides. EMBO J 1987;6:1245–1249.

44 Madden DR, Gorga JC, Strominger JL, Wiley DC: The structure of HLA-B27 reveals nonamer self-peptides bound in an extended conformation. Nature 1991;353:321–325.

45 Brown JH, Jardetzky T, Saper MA, Samraoui B, Bjorkman PJ, Wiley DC: A hypothetical model of the foreign antigen binding site of class II histocompatibility antigens. Nature 1988;332:845–848.

46 Guillet J-G, Lai M-Z, Briner TJ, Buus S, Sette A, Grey HM, Smith JA, Gefter ML: Immunological self non-self discrimination. Science 1987;235:865–870.

47 Rothbard JB, Lechler RI, Howland K, Bal V, Eckels DD, Sekaly R, Long EO, Taylor WR, Lamb JR: Structural model of HLA-DR1 restricted T cell antigen recognition. Cell 1988;52:512–523.

48 Manca F, Habeshaw J, Dalgleish A: The naive repertoire of human helper cells specific for gp120, the envelope glycoprotein of HIV. J Immunol 1991;146:1964–1971.

49 Abrignani S, Montagna D, Jeannet M, Wintsch J, Haigwood NL, Shuster JR, Steimer KS, Cruchaud A, Stahelin T: Priming of $CD4^+$ T cells specific for conserved regions of human immunodeficiency virus glycoprotein gp120 in humans immunized with a recombinant envelope protein. Proc Natl Acad Sci USA 1990;87: 6136–6140.

50 Botarelli P, Houlden BA, Haigwood NL, Servis C, Montagna D, Abrignani S: N-glycosylation of HIV-gp120 may constrain recognition by T lymphocytes. J Immunol 1991;147:3128–3132.

51 Wain-Hobson S, Varanian J-P, Henry M, Chenciner N, Cheynier R, Delassus S, Martins LP, Sala M, Nugeyre M-T, Guetard XD, Klatzmann D, Gluckman J-C, Rozenbaum W, Barré-Sinoussi F, Montagnier L: LAV revisited: Origins of the early HIV-1 isolates from Institut Pasteur. Science 1991;252:961–965.

52 Gallo RC: Nature 1991;351:358.

53 Lasky LA, Nakamura G, Smith DH, Fennie C, Shimasaki C, Patzer E, Berman P, Gregory T, Capon DJ: Delineation of a region of the human immunodeficiency virus type 1 gp120 glycoprotein critical for the interaction with the CD4 receptor. Cell 1987;50:975–985.

54 De Groot AS, Clerici M, Hosmalin A, Hughes SH, Barnd D, Hendrix CW, Houghten R, Shearer GM, Berzofsky JA: Human immunodeficiency virus reverse transcriptase T helper epitopes identified in mice and humans: Correlation with a cytotoxic T cell epitope. J Infect Dis 1991;164:1058–1065.

55 Krowka J, Stites D, Debs R, Larsen C, Fedor J, Brunette E, Duzgunes N: Lymphocyte proliferative responses to soluble and liposome-conjugated envelope peptides of HIV-1. J Immunol 1990;144:2535–2540.

56 Takahashi H, Germain RN, Moss B, Berzofsky JA: An immunodominant class I-restricted cytotoxic T lymphocyte determinant of human immunodeficiency virus type 1 induces CD4 class II-restricted help for itself. J Exp Med 1990;171:571–576.
57 Haas G, David R, Frank R, Gausepohl H, Devaux C, Claverie J-M, Pierres M: Identification of a major human immunodeficiency virus-1 reverse transcriptase epitope recognized by mouse CD4+ T lymphocytes. Eur J Immunol 1991;21:1371–1377.
58 Palker TJ, Matthews TJ, Langlois A, Tanner ME, Martin ME, Scearce RM, Kime JE, Berzofsky JA, Bolognesi DP, Haynes BF: Polyvalent human immunodeficiency virus synthetic immunogen comprised of envelope gp120 T helper cell sites and B cell neutralization epitopes. J Immunol 1989;142:3612–3619.
59 Van Bleek GM, Nathenson SG: Isolation of an endogenously processed immunodominant viral peptide from the class I H-2K^b molecule. Nature 1990;348:213–216.
60 Jardetzsky TS, Lane WS, Robinson RA, Madden DR, Wiley DC: Identification of self peptides bound to purified HLA-B27. Nature 1991;353:326–329.
61 Rudensky AY, Preston-Hurlburt P, Hong S-C, Barlow A, Janeway CA Jr: Sequence analysis of peptides bound to MHC class II molecules. Nature 1991;353:622–627.
62 Henderson RA, Michel H, Sakaguchi K, Shabanowitz J, Appella E, Hunt DF, Englehard V: HLA-A2.1-associated peptides from a mutant cell line: A second pathway of antigen presentation. Science 1990;255:1264–1266.
63 Falk K, Rotzschke O, Deres K, Metzger J, Jung G, Rammensee H-G: Identification of naturally processed viral nonapeptides allows their quantification in infected cells and suggests an allele-specific T cell epitope forecast. J Exp Med 1991;174:425–434.
64 Rotzschke O, Falk K, Faath S, Rammensee H-G: On the nature of peptides involved in T cell alloreactivity. J Exp Med 1991;174:1059–1071.
65 Falk K, Rotzschke O, Stevanovic S, Jung G, Rammensee H-G: Allele-specific motifs revealed by sequencing of self peptides eluted from MHC molecules. Nature 1991; 351:290–296.

Robert F. Siliciano, Department of Medicine, Johns Hopkins University
School of Medicine, Baltimore, MD 21205 (USA)

Norrby E (ed): Immunochemistry of AIDS.
Chem Immunol. Basel, Karger, 1993, vol 56, pp 150–164

How Does the HIV Escape Cytotoxic T Cell Immunity?

R.E. Phillips, A.J. McMichael

Institute of Molecular Medicine, Molecular Immunology Group, John Radcliffe Hospital, Headington, Oxford, UK

A cardinal feature of human immunodeficiency virus (HIV) infection is its long but variable latent period. If the mechanisms responsible for the suppression of overt manifestations of infection were understood, then perhaps this latent state could be prolonged indefinitely with appropriate intervention.

The life cycle of a retrovirus offers opportunities to establish infection with proviral DNA which is not transcriptionally active. It is likely that failure to produce viral proteins will protect the retrovirus from detection by the immune system. Two models have been proposed based on experiments in cell culture with HIV-1. In the first model, the HIV virion enters a quiescent T lymphocyte but reverse transcription and integration of the resultant double-stranded proviral intermediate is inefficient [1]. Evidence to support this idea comes from in vitro experiments which suggest that extra chromosomal HIV-1 genomes persist in infected cells until those cells are activated [1]. Further evidence for this idea includes the demonstration of high amounts of unintegrated HIV-1 proviruses in T cells recovered from infected individuals [1].

A second molecular mechanism proposed to account for HIV latency involves integration of the provirus into the genome of non-replicating cells. In this model, the provirus remains quiescent until an external stimulus causes production of the viral transactivating protein, tat, which is involved in promoting the synthesis of viral mRNA transcripts which encode structural proteins [2]. Some cell cultures chronically infected with HIV, such as

U-1 or ACH-2, contain little of the mRNA encoding gag and pol until stimulation with 12-O-tetradecanocyl phorbol 13-acetate or tumour necrosis factor-α [3]. Although transcription of HIV proviral templates does appear to be suppressed in these cell cultures, small amounts of mature virions are still produced [4]. In any case, specially adapted cells in culture infected with laboratory strains of virus are very unlikely to reflect the biology of infections in vivo where some cells may be transcriptionally active while others are quiescent.

Clinical studies show quite clearly that viral replication continues during the asymptomatic phase of infection. Ho et al. [5] recovered HIV from the plasma and from the peripheral-blood monocytes of all 54 seropositive individuals they studied. In a similar study, Coombs et al. [6], isolated HIV from 207 of 213 seropositive individuals. These findings, as well as the molecular data which demonstrate a continuous evolution of sequence change in the virus [7, 8], indicate that HIV should be regarded as a persistent infection characterized by continuous viral replication, which is responsible for generating considerable genetic diversity.

Throughout the asymptomatic phase of infection, there is a specific antibody and T cell response to the virus which is readily demonstrated. This finding is consistent with the idea that continuous replication of virus constantly stimulates specific immune responses. Why is this immunity unable to clear the virus? Investigation of this problem requires a detailed understanding of antiviral immunity in HIV. If the molecular basis of this immune response was known, we could then hope to understand how the virus might subvert this defense mechanism.

Immunity to Viruses

All viruses that infect humans elicit both antibody- and cell-mediated responses. Antibody is able to neutralize virus in vitro through binding to attachment or fusion sites on viral glycoproteins [9, 10]. In some viruses mutation within these glycoproteins allows virus to escape this neutralization. This phenomenon is understood at a molecular level for influenza hemagglutinin [11]. It is a major feature of the gp120 protein of HIV [12, 13]. Such escape mutation is usually taken as evidence that the antibody is effective. Therefore, evasion of antibody responses is very likely to play a part in the persistence of HIV in vivo, although this has not been formally proven. The convergence of gp120 V3 loop sequences in early HIV infection followed by divergence, seen by Holmes et al. [14] and Goudsmit et al. [pers. communication] is further evidence that selection forces, probably antibody-mediated, are responsible for these genetic shifts.

In experimental animals, there is strong evidence that cytotoxic T lymphocytes (CTLs) are also an important component of the immune response to viruses [15]. Transfer of CD8+ spleen cells from mice infected with influenza cleared virus from the lungs and prolonged survival of infected recipients [16, 17]. H2bm1 mice which are unable to mount a CTL response to Sendai virus die, whereas H2b mice mount a specific CTL immunity and survive [18]. Early work also suggested that CTLs protected against oncogenic retroviruses [19–21].

Passive transfer experiments showed that mice could be protected against murine sarcoma virus-induced tumours [22] and against Friend murine leukemia virus-induced leukemia, by either CD8 or CD4 T lymphocytes primed against the appropriate retrovirus in vivo [23]. CD8+ CTLs also protect naive mice against challenge by a retrovirus.

In humans, CTL levels correlate with protection against influenza and respiratory syncytial virus infection in volunteers [24, 25]. In persistent infections with Epstein-Barr virus (EBV) and cytomegalovirus, CTLs are activated and probably play a major role in containing the virus without eliminating it [26–28]. In SCID mice reconstituted with human peripheral-blood lymphocytes, EBV-positive lymphomas commonly develop, probably because of impaired T cell immunity under these conditions [27]. In vitro, CTLs clearly regulate EBV infection and transformation of B lymphocytes; specific CTLs caused regression of a B cell line [26]. Similarly, CTLs have been shown to suppress the replication of both HIV and simian immunodeficiency viruses in vitro [29–31].

Although there are compelling reasons to think that CTLs might perform an important and perhaps critical role in control of HIV infection, several questions arise. Firstly, if as has been proposed, HIV must be regarded throughout the asymptomatic period as an infection in which virus is continually produced, why are infected cells not destroyed and the virus contained? Second, if the virus does sequester at least transiently where replication can proceed unchecked to produce antigenic variants, why do these variants not elicit a novel immune response which then clears the virus? In this review, we propose to examine the evidence that HIV is able to escape the CTL response that it so readily elicits. To do so, we will first review the molecular mechanisms through which CTLs recognize viruses.

Cytotoxic T Cell Recognition of Viruses

Over the last 17 years, there has been considerable progress made towards elucidating how T lymphocytes are able to recognize intracellular organisms [32–35]. The essential feature of this system is the transport of viral antigen to the cell surface in a form which can be recognized by the T cell antigen receptor [36–38].

In 1974, Zinkernagel and Doherty [32] showed that murine CTLs lysed target cells infected with lymphocytic choriomeningitis virus (LCMV) only if these cells shared the major histocompatibility complex (MHC) antigen, H2. One interpretation of this discovery was that CTLs recognized a combination of viral protein and the class I molecule. Although this hypothesis has proved to be correct, early models demanded that whole viral surface proteins appeared on the cell surface in close proximity to the class I molecule. In a series of experiments, Townsend and others refuted this idea. CTL clones capable of recognizing the internal nucleoprotein of influenza could be isolated [39, 40]. In further experiments it was shown that CTLs could recognize and lyse target cells which contained no more than truncated fragments of influenza nucleoprotein which lacked any sequences capable of signalling intracellular transport [34]. In these experiments, some protein fragments failed to act as CTL targets, suggesting that the CTLs recognized certain specific portions of the viral protein. This was proved when short synthetic peptides derived from viral protein sequences were shown to be sufficient to act as antigens for CTLs [36]. In humans, similar results were obtained and it became possible to locate the CTL antigenic sites (epitopes) within a viral protein using overlapping synthetic peptides [41–43].

When CTL epitopes were mapped with the use of these synthetic peptides, it became increasingly clear that the HLA class I molecule dictated where, within a viral protein, an epitope would be selected. Individuals who utilized the same class I molecule to mount a CTL response to a viral protein recognized the same peptide fragment. Why this should be so became clearer when the crystal structures of HLA-A2 [44–46] and later HLA-A68 [47] and HLA-B27 [48] were solved. The class I molecule consisted of two α-helices with a floor lying between them formed by a β-pleated sheet. The α-helices bordered a cleft; most of the considerable polymorphism of the class I molecules involved amino acids which either pointed into this cleft from these helices, or from the sheet below. In the original crystal structures, electron-dense material was detected within the cleft and we now know that this is the site where peptide antigens bind. HLA polymorphism alters the shape of the antigen-binding cleft and so in turn determines to a substantial degree which peptides can bind [49, 50]. These structural studies also provided a direct demonstration of the sort of configuration that an HLA class I molecule containing a viral peptide would display to the antigen receptor of a cytotoxic T cell.

Antigen Processing and HLA Class I Binding

Another technical development which has revolutionized this field has been the direct isolation and sequencing of naturally occurring peptides. Using acid extraction techniques it has been possible to isolate peptides from the cellular pool and from precipitated HLA molecules [49, 50]. This work has further verified the idea that short peptides sit in the HLA binding groove. Contrary to earlier predictions, naturally occurring peptides are usually no more than nine amino acids in length. When the pool of eluted peptides is sequenced a heterogeneous pattern is obtained, but at certain positions along the peptide length there is a high degree of sequence conservation. For example, in the 9-amino-acid peptide eluted from HLA-A2, position 2 is usually occupied by leucine, while at position 9 there is a hydrophobic carboxy-terminal residue which is usually valine or leucine [49, 51]. Similar conservation at certain positions has been detected in peptides eluted from other class I molecules. When these peptide sequences are modeled into the HLA binding cleft, it is possible to infer with a high measure of confidence that conserved amino acids are involved in critical binding or anchoring to pockets on the surface of the cleft [49], which in crystal studies have been shown to contain electron-dense material. This work accords very closely with the concept that the class I molecule is the principal factor in determining which viral peptides can bind and so become antigenic targets. The elegance of this system also reveals a potential weakness: if only certain peptides can bind and allow class I molecules to present them on the cell surface, then any mutation within the epitope which prevents binding might lead to a loss of immune recognition. The structural requirements for recognition of the peptide-class I complex by the T cell receptor are much less well understood, but it is likely that some amino acid substitutions which did not prevent peptide binding could alter the configuration presented to the T cell receptor.

There has recently been rapid progress in elucidating how viral proteins are degraded and transported to the site of class I synthesis in the endoplasmic reticulum of the cell. Much of the stimulus for this work has come from the study of mutant cell lines which have defects in antigen presentation. Townsend et al. [36] showed that the murine lymphoma mutant RMA-S had a defect in the assembly and presentation of MHC class I molecules which could be corrected by supplying large quantities of an appropriate synthetic viral peptide in the extracellular fluid. A similar result was obtained with the mutant human B lymphoblastoid cell line 721.174 (and its derivative T2) [52]. Mapping experiments showed that 721.174 and T2 cells had a large deletion in the MHC complex which was presumably responsible for the phenotype of the cell. Since the class I molecules appeared to be normally synthesized in the endoplasmic reticulum, it was hypothesized that the mutant cells lacked genes which generate and transport peptides from the cytoplasm (where degradation is thought to occur) to the endoplasmic reticulum where class I molecules are synthesized. In more recent work, particular genes cloned from the class II region of the MHC in the deleted 721.174 cells have been shown to correct the failure of class I assembly in some of the mutant cell lines [53, 54]. Since these genes encode proteins which have strong homology to transporters, these complementation experiments confirm Townsend et al.'s initial predictions. It is also clear that there is polymorphism in these transporter proteins which may influence the selection of peptides which are delivered to nascent class I molecules [54].

Other genes in the same region of the MHC encode two proteins that physically associate with an approximately 750-kDa complex made up of twenty or more subunits, known as the proteasome [55–57]. These MHC gene products are known as the LMP particle (for low-molecular-weight polypeptides). The proteasome is the leading candidate for the site of proteolytic degradation of viral proteins, although this hypothesis awaits experimental confirmation. LMP proteins show some polymorphism; they are closely linked to genes encoding the putative peptide transporter and are induced by γ-interferon. These features provide a strong impetus to investigate the role of the proteasome in antigen presentation and may provide insight into ways that viral mutation might subvert these preliminary stages of immune recognition.

Cytotoxic T Cells Specific for HIV

It is now clear that HIV elicits a specific CTL response in the vast majority of asymptomatic individuals. Work in several laboratories has shown that HIV-specific CTLs are capable of recognizing most of the structural and regulatory proteins of the virus [reviewed in ref. 58]. Precursor frequencies are often high [59, 60] and, in some asymptomatic individuals, specific lysis of target cells expressing HIV antigens can be elicited without preliminary lymphocyte culture [61–63]. As with other human viruses, it has been possible to map the sites of CTL recognition within HIV proteins using sets of overlapping synthetic peptides [64]. In initial studies we defined an HLA-B27-restricted epitope in HIV gag using a peptide designated p24-14 (gag 265–279) [64]. This response was sustained during serial assays, and in other individuals with HLA-B27 the same peptide is recognized. By contrast, the HLA-B8-restricted response to HIV gag was mapped with three distinct peptides, p17-3 (gag 24–35), p24-13 (gag 255–269) and p24-20 (gag 325–339). When responses to these peptide epitopes were followed over time, fluctuating patterns of recognition were found. At a given time, CTLs might recognize only one or two of the HLA-B8-restricted HIV gag epitopes. As an individual was followed, a predominant response might wane and recognition of the second or third B8-restricted epitope might appear. Why should this be so?

Genetic variation is a well-known characteristic of retroviruses. As isolates of HIV were first sequenced, considerable variation was apparent, but the extent of diversity within individual isolates became clear with the use of the polymerase chain reaction. These studies showed that variation was most marked in the envelope gene of HIV [65] while the structural genes (gag and pol) were much more conserved. Sequence conservation in these genes presumably reflects the lethal consequences of mutation which might disrupt the function of reverse transcriptase (RT), protease or integrase or the structure of gag. However, detailed analysis of variation in pol and gag has shown that some genetic variation productive of amino acid substitutions is tolerated by the virus. This is best illustrated by mutations in HIV RT which confer resistance to zidovudine in vitro [66, 67]. Sequential studies of isolates from individuals treated with this drug have clearly demonstrated the acquisition of multiple amino acid substitutions, which do not disturb the function of the enzyme.

The rate of genetic variation in HIV arises from three variables: mutation rate per replication cycle, number of replication cycles per unit time and the selective advantage or disadvantage possessed by the variant. Overall estimates of rates of change in HIV are not feasible, although the misincorporation rate of HIV RT is thought to be in the order of 1 in 10,000 base pairs [68]. Comparative estimates suggest that the error rate of HIV RT is higher than that of other retroviruses [68]. It is usual to blame the lack of proofreading capacity of RT for mutation of the virus, but transcription of the DNA provirus by RNA polymerase 2 is another potential source of error in the life cycle of the virus. Intragenomic rearrangements such as deletions, duplications, inversions or combinations of these also increase the genetic diversity of retroviruses.

Genetic Variation in the Cytotoxic Epitopes of HIV gag

The discovery that CTL responses to three HLA-B8-restricted epitopes fluctuated, raised the question as to whether genetic variation in the virus might be in some way responsible [69]. It was well known that genetic variation in the V3 loop of the HIV envelope could lead to loss of neutralization of virus by antibody in vitro [9, 10], but there had been no clear demonstration of CTL escape mutants in any naturally occurring viral infection. Pircher et al. [70] found LCMV variants which escaped CTL recognition by acquiring mutation within the gene fragment encoding the epitope. However, in this experiment, mice were transgenic for a single T cell receptor and the vast majority of circulating T cells expressed the transgenic product. This work and a subsequent experiment in which cloned CTLs selected mutant LCMV in vitro [71] did provide support for the idea that CTLs could exert selection pressure on viruses in vivo and suggested conditions that might have to be present before this could be detected. First, the CTL response should be directed towards a region of the virus which could retain function despite amino acid substitutions. Second, the CTL response might need to be directed towards a few epitopes. Finally, target cells would need to be infected with mutant virus alone, since epitopes arising from wild-type virus would sensitize the infected cell for lysis whether a mutant epitope was present or not.

In initial studies, we sequenced the HIV proviral DNA extracted from the peripheral-blood mononuclear cells taken from individuals who recognized the HLA-B27-restricted epitope p24-14. We did find one amino acid substitution of position 270 in most isolates (leucine to methionine) in our proband patient 007, and in two others. When a synthetic peptide based on this variant sequence was tested 007 CTLs invariably lysed target cells more efficiently [Rowland-Jones et al., unpubl. data]. Indeed, one CTL clone was isolated that recognized this variant peptide and not the published sequence. A similar study [8], found more extensive sequence variation in p24-14 and tested several of the peptides but not all: these variants sensitized target cells as effectively as the prototype sequence. However, one of the variants (glycine to glutamic acid at residue 271) detected in that study was made by us in Oxford, and failed to be recognized by CTLs from an HLA-B27 donor [Rowland-Jones et al., unpubl. data]. The great degree of variability in Meyerhans et al.'s study [8] may reflect the amplification of unintegrated defective proviruses in resting peripheral-blood lymphocytes (e.g. some sequences contain stop codons) compared to our study where many (but not all) of the sequences were derived from cultured cells and were probably integrated and replicative, and therefore functional, sequences.

There was considerable genetic variation in regions of HIV gag encoding HLA-B8-restricted epitopes. Much, but not all, of this variation was within the epitopes themselves. When synthetic peptides based on variation in p24-13 gag were tested, fluctuating responses were seen. The E (glutamic acid) → D (aspartic acid) change at position 262 seemed to be critical. Many of the sequences in both patients analyzed showed the 'D' variant; all CTL cultures recognized the peptide representing this sequence. Later sequences contained an increasing proportion of the E version and some CTL clones at later time points failed to recognize this sequence. This finding suggests that as new epitope sequences evolve, a new T cell response is elicited, but that some changes may result in epitopes not seen by all CTLs.

Variation in amino acid sequences encoding an HLA-B8-restricted epitope in p17 (17-3) was found in two patients. By comparison, we found no sequence variation in three other HLA-B8-negative patients who did not recognize any CTL antigenic target in p17 gag. This finding strongly suggested that variation in p17 was largely driven by immune selection. Sequence variation was also seen in p17 but outside the epitope-coding region in both HLA-B8 donors but not in the HLA-B27 controls. This variation thus appeared to be linked to HLA class I immune recognition, but the sites where this variation occurred did not appear to be cytotoxic T cell epitopes since no peptide sequence based on this variation was found to sensitize target cells in lysis assays. It is possible that mutation outside the p17-3 epitope region might affect antigen processing [72–74] or alter the capacity of polymorphic transporter molecules to deliver variant polypeptides to the endoplasmic reticulum. If this were so, then variation in p17 of this sort which reduced the efficiency of epitope formation could confer a survival advantage to the virus as readily as epitope changes. In fact, any viral mutation which impaired HLA class I immune recognition might have this effect.

When synthetic peptides based on the naturally occurring variation in p17-3 were tested in lysis assays several findings were obtained. In both donors the p17-3 epitope had been mapped with a 15-amino-acid peptide derived from the reference sequence HIV-SF2. However, within this region the sequence variation we detected lay exclusively in the last ten amino acids. At position 26 a substitution of K→R (lysine to arginine) completely abolished CTL recognition in both donors who had this sequence detected in peripheral-blood lymphocyte DNA. CTLs obtained from these donors at several points in time also failed to recognize this variant peptide and a second form which in addition had an L→F (leucine to phenylalanine) substitution at position 31. It is possible that these variant peptides are not recognized by CTLs because they fail to bind to HLA-B8 and so never appear on the target cell surface. Interestingly, the gag sequence with arginine at position 26 appears to be increasing in frequency in the blood of at least one of the patients [Edwards et al., unpubl. observation].

Conclusion

This work provides experimental evidence that immune selection influences the diversity of HIV found during the natural course of infection. It is not possible to weigh the relative importance of different arms of the immune system as a means of controlling HIV during the long asymptomatic period. However, the appearance of sequence variation which alters the ability of cytotoxic T cells to recognize antigens of the virus is good evidence that this form of immunity exerts a selective force. We are still profoundly ignorant as to how the virus persists and ultimately overwhelms the immune system. The detection of 'escape' forms of antigen suggests one way that the virus might acquire a survival advantage. If this is so, then antigenic variation might also be detected in epitopes within highly conserved genetic loci, such as RT, and our preliminary results suggest that this is so [Phillips et al., unpubl. data]. It is also conceivable that CTL-driven antigenic diversity will increase over time and that progression of the infection might rely on the appearance of a widening range of viral quasispecies. In this model, the capacity of the T cell repertoire to cope with viral mutation is taxed while some antigenic variants are unseen through subversion of the HLA class I system. These ideas are testable and also provide a basis for investigating the limits of the immune system's ability to contain a highly mutable virus.

References

1 Stevenson M, Stanwick TL, Dempsey MP, Lamonica CA: HIV-1 replication is controlled at the level of T cell activation and proviral integration. EMBO J 1990;9: 1551–1560.

2 Pomerantz RJ, Trono D, Feinberg MB, Baltimore D: Cells nonproductively infected with HIV-1 exhibit an aberrant pattern of viral RNA expression: A molecular model for latency. Cell 1990;61:1271–1276.

3 Griffin GE, Leung K, Folks TM, Kunkel S, Nabel GJ: Activation of HIV gene expression during monocyte differentiation by induction of NF-kappa B. Nature 1989; 339:70–73.

4 Folks TM, Justement J, Kinter A, et al: Characterization of a promonocyte clone chronically infected with HIV and inducible by 13-phorbol-12-myristate acetate. J Immunol 1988;140:1117–1122.

5 Ho DD, Moudgil T, Alam M: Quantitation of human immunodeficiency virus type 1 in the blood of infected persons. N Engl J Med 1989;321:1621–1625.

6 Coombs RW, Collier AC, Allain JP, et al: Plasma viremia in human immunodeficiency virus infection. N Engl J Med 1989;321:1626–1631.

7 Simmonds P, Balfe P, Peutherer JF, Ludlam CA, Bishop JO, Brown AJ: Human immunodeficiency virus-infected individuals contain provirus in small numbers of peripheral mononuclear cells and at low copy numbers. J Virol 1990;64:864–872.

8 Meyerhans A, Dadaglio G, Vartanian J-P, et al: In vivo persistence of a HIV-1-encoded HLA B27 restricted cytotoxic T lymphocyte epitope despite specific in vitro reactivity. Eur J Immunol 1991;21:2637–2640.

9 Looney DJ, Fisher AG, Putney SD, et al: Type-restricted neutralization of molecular clones of human immunodeficiency virus. Science 1988;241:357–359.

10 Albert J, Naucler A, Bottiger B, et al: Replicative capacity of HIV-2, like HIV-1, correlates with severity of immunodeficiency. AIDS 1990;4:291–295.

11 Wiley DC, Wilson IA, Skehel JJ: Structural identification of the antibody-binding sites of Hong Kong influenza haemagglutinin and their involvement in antigenic variation. Nature 1981;289:373–378.

12 Reitz MJ, Wilson C, Naugle C, Gallo RC, Robert GM: Generation of a neutralization-resistant variant of HIV-1 is due to selection for a point mutation in the envelope gene. Cell 1988;54:57–63.

13 LaRosa GJ, Davide JP, Weinhold K, et al: Conserved sequence and structural elements in the HIV-1 principal neutralizing determinant. Science 1990;249:932–935.

14 Holmes EC, Zhang LQ, Simmonds P, Ludlam CA, Leigh-Brown AJ: Convergent and divergent sequence evolution in the surface envelope glycoprotein of human immuno deficiency virus type I within a single infected patient. Proc Natl Acad Sci USA, in press.

15 Bangham CRM, McMichael AJ: T-cell immunity to viruses; in Feldmann M, Lamb J, Owen J (eds): T Cells. New York, Wiley, 1989, pp 281–310.

16 Lin Y, Askonas BA: Biological properties of an influenza A virus specific killer T cell clone. J Exp Med 1981;154:225–234.

17 Yap KL, Ada GL: Transfer of specific cytotoxic T lymphocytes protects mice inoculated with influenza virus. Nature 1978;273:238–240.

18 Kast WM, Bronkhorst AM, de Waal LP, Melief CJ: Cooperation between cytotoxic and helper T lymphocytes is protective against lethal Sendai virus infection. J Exp Med 1986;164:723–738.
19 Blank KJ, Freedman HA, Lilly P: T-lymphocyte response to Friend virus-induced tumor cell-lines in mice congenic at H-2. Nature 1976;260:250–252.
20 Collavo D, Ronchese F, Zanovello P, Biasi G, Chieco-Bianchi L: T cell tolerance in Moloney-murine leukemia (M-MuLV) carrier mice: Low cytotoxic T lymphocyte precursor frequency and absence of suppressor T cells in carrier mice with Moloney-murine sarcoma (M-MSV)-induced tumours. J Immunol 1982;128:774–783.
21 Plata F, Cerottini JC, Brunner KT: Primary and secondary in vitro generation of cytolytic T lymphocytes in the murine sarcoma virus system. Eur J Immunol 1975;5: 227–233.
22 Leclerc JC, Gomard E, Plata F, Levy J-P: Cell-mediated immune reaction against tumours induced by oncornaviruses. II. Nature of effector cells in tumour-cell cytolysis. Int J Cancer 1973;11:426–432.
23 Greenberg PD, Cheever MA, Fefer A: Eradication of disseminated murine leukemia by chemoimmunotherapy with cyclophosphamide and adoptively transferred immune syngeneic Lyt-1+2– lymphocytes. J Exp Med 1981;154;952–963.
24 McMichael AJ, Gotch FM, Noble GR, Beare PAS: Cytotoxic T-cell immunity to influenza. N Engl J Med 1983;309:13–17.
25 Isaacs D, McDonald NE, Bangham CRM, McMichael AJ, Higgins PG, Tyrrell DAJ: The specific cytotoxic T-cell response of adult volunteers to infection with respiratory syncytial virus. Immun Infect, in press.
26 Moss DJ, Rickinson AB, Pope JH: Long term T cell mediated immunity to Epstein-Barr virus in man. I. Complete regression of virus-induced transformation on cultures of seropositive donor leukocytes. Int J Cancer 1978;22:662–668.
27 Rowe M, Young LS, Crocker J, Stokes H, Henderson S, Rickinson AB: Epstein-Barr virus (EBV)-associated lymphoproliferative disease in the SCID mouse model: Implications for the pathogenesis of EBV-positive lymphomas in man. J Exp Med 1991;173:147–158.
28 Quinnan GV, Kirmani N, Rook AH, et al: Cytotoxic T cell in cytomegalovirus infection: HLA restricted T lymphocyte and non-T lymphocyte cytotoxic responses correlate with recovery from cytomegalovirus infection on bone-marrow transplant recipients. N Engl J Med 1982;307:7–13.
29 Walker CM, Moody DJ, Stites DP, Levy JA: CD8+ lymphocytes can control HIV infection in vitro by suppressing virus replication. Science 1986;234:1563–1566.
30 Kannagi M, Masuda T, Hattori T, et al: Interference with human immunodeficiency virus (HIV) replication by CD8+ T cells in peripheral blood leukocytes of asymptomatic HIV carriers in vitro. J Immunol 1990;64:3399–3406.
31 Brinchmann JE, Gaudernack G, Vartdal F: CD8+ T cells inhibit HIV replication in naturally infected CD4+ T cells: Evidence for a soluble inhibitor. J Immunol 1990; 144:2961–2966.
32 Zinkernagel RM, Doherty PC: H-2 compatibility requirement for T-cell mediated lysis of target cells infected with lymphocytic choriomeningitis virus. J Exp Med 1975;141:1427–1436.
33 Townsend ARM, McMichael AJ, Cartner NP, Huddleston JA, Brownlee GG: Cytotoxic T cell recognition of the influenza nucleoprotein and haemagglutinin expressed in transfected mouse L cells. Cell 1984;39:13–25.

34 Townsend ARM, Gotch FM, Davey J: Cytotoxic T cells recognise fragments of influenza nucleoprotein. Cell 1985;42:457–467.

35 Townsend A, Rothbard J, Gotch F, Bahadur B, Wraith D, McMichael A: The epitopes of influenza nucleoprotein recognized by cytotoxic T lymphocytes can be defined with short synthetic peptides. Cell 1986;44:959–968.

36 Townsend A, Bastin J, Gould K, et al: Defective presentation to class I restricted cytotoxic T lymphocytes in vaccinia infected cells is overcome by enhanced degradation of antigen. J Exp Med 1988;168:1211–1224.

37 Townsend A, Ohlen C, Foster L, Bastin J, Ljunggren HG, Karre K: A mutant cell in which association of class I heavy and light chains is induced by viral peptides. Cold Spring Harb Symp Quant Biol 1989;1:299–308.

38 Townsend A, Bodmer H: Antigen recognition by class I-restricted cytotoxic T lymphocytes. Annu Rev Immunol 1989;7:601–624.

39 Bennink JR, Yewdell JW, Gerhard W: A viral polymerase involved in recognition of influenza virus-infected cells by a cytotoxic T cell clone. Nature 1982;296:75–76.

40 Townsend ARM, Skehel JJ: Influenza A specific cytotoxic T cell clones that do not recognise viral glycoproteins. Nature 1982;300:655.

41 McMichael AJ, Michie CA, Gotch FM, Smith GL, Moss B: Recognition of influenza A nucleoprotein by human cytotoxic T lymphocytes. J Gen Virol 1986;67:719–726.

42 McMichael AJ, Gotch FM, Santos-Aguado J, Strominger JL: Effect of mutations and variations of HLA-A2 on recognition of a virus peptide epitope by cytotoxic T lymphocytes. Proc Natl Acad Sci USA 1988;85:9194–10006.

43 Gotch F, Rothbard J, Howland K, Townsend A, McMichael A: Cytotoxic T lymphocytes recognise a fragment of influenza virus matrix protein in association with HLA-A2. Nature 1987;326:881–882.

44 Bjorkman P, Saper M, Samraoui B, Bennett W, Strominger J, Wiley D: Structure of human class I histocompatibility antigen, HLA-A2. Nature 1987;329:506–511.

45 Bjorkman P, Saper M, Samraoui B, Bennett W, Strominger J, Wiley D: The foreign antigen binding site and T cell recognition regions of class I histocompatibility antigens. Nature 1987;329:512–519.

46 Saper MA, Bjorkman PJ, Wiley DC: Refined structure of the human histocompatibility antigen HLA-A2 at 2.6 Å resolution. J Mol Biol, in press.

47 Garrett TPJ, Saper MA, Bjorkman PJ, Strominger JL, Wiley DC: Specificity pockets for the side chains of peptide antigens in HLA-Aw68. Nature 1989;342:692.

48 Madden DR, Gorga JC, Strominger JL, Wiley DC: The structure of HLA-B27 reveals nonamer 'self-peptides' bound in an extended conformation. Nature 1991;353:321–325.

49 Falk K, Rotzschke O, Stevanovic S, Jung G, Rammensee H-G: Allele specific motifs revealed by sequencing of self peptides eluted from MHC molecules. Nature 1991; 351:290–296.

50 Jardetsky TS, Lane WS, Robinson RA, Madden DR, Wiley DC: Identification of self peptides bound to purified HLA-B27. Nature 1991;353:326–329.

51 Hunt DF, Henderson RA, Shabanowitz J, et al: Characterization of peptides bound to the class I MHC molecule HLA A2.1 by mass spectrometry. Science 1992;255: 1261–1263.

52 Cerundolo V, Alexander J, Anderson K, et al: Presentation of viral antigen controlled by a gene in the major histocompatibility complex. Nature 1990;345:449–452.

53 Spies T, DeMars R: Restored expression of major histocompatibility class I molecules by gene transfer of a putative peptide transporter. Nature 1991;351:323–325.
54 Powis SJ, Deverson EV, Coadwell WJ, et al: Effect of polymorphism of an MHC-linked transporter on the peptides assembled in a class I molecule. Nature 1992;357: 207–215.
55 Monaco JJ, McDevitt HO: H-2-linked low-molecular weight polypeptide antigens assemble into an unusual macromolecular complex. Nature 1984;309:797–799.
56 Brown MG, Driscoll J, Monaco JJ: Structural and serological similarity of MHC-linked LMP and proteosome (multicatalytic proteinase) complexes. Nature 1991; 353:365–367.
57 Martinez CK, Monaco JJ: Homology of the proteasome subunits to a major histocompatibility complex-linked LMP gene. Nature 1991;353:664–667.
58 Nixon DF, McMichael AJ: Cytotoxic T cell recognition of HIV proteins and peptides. AIDS 1991;5:1049–1059.
59 Gotch FM, Nixon DF, Alp N, McMichael AJ, Borysiewicz LK: High frequency of memory and effector gag specific cytotoxic T lymphocytes in HIV seropositive individuals. Int Immunol 1990;2:707–712.
60 Hoffenbach A, Langlade-Demoyen P, Dadaglio G, et al: Unusually high frequencies of HIV-specific cytotoxic T lymphocytes in humans. J Immunol 1989;142:452–462.
61 Plata F, Autran B, Martins LP, et al: AIDS virus specific cytotoxic T lymphocytes in lung disorders. Nature 1987;328:348–351.
62 Walker BD, Flexner C, Paradis TJ, et al: HIV-1 reverse transcriptase is a target for cytotoxic T lymphocytes in infected individuals. Science 1988;240:64–66.
63 Walker DB, Chakrabati S, Moss B, et al: HIV specific cytotoxic T lymphocytes in seropositive individuals. Nature 1987;328:345–348.
64 Nixon DF, Townsend ARM, Elvin JG, Rizza CR, Gallwey J, McMichael AJ: HIV-1 gag-specific cytotoxic T lymphocytes defined with recombinant vaccinia virus and synthetic peptides. Nature 1988;336:484–487.
65 Hahn BH, Shaw GM, Taylor ME, et al: Genetic variation in HTLV-III/LAV over time in patients with AIDS or at risk for AIDS. Science 1986;232:1548–1553.
66 Larder BA, Kemp SD, Purifoy DJM: Infectious potential of human immunodeficiency virus type 1 reverse transcriptase mutants with altered inhibitor sensitivity. Proc Natl Acad Sci USA 1989;86:4803–4807.
67 Larder BA, Kellam P, Kemp SD: Zidovudine resistance predicted by direct detection of mutations in DNA from HIV-infected lymphocytes. AIDS 1991;5:137–144.
68 Coffin JM: Genetic variation in retroviruses; in Kurstak E, Marusk RG, Murphy FA, VanRegenmortel MHV (eds): Virus Variability, Epidemiology, and Control; in Marusk RG, Murphy FA, VanRegenmortel MHV (eds): Applied Virology Research. New York, Plenum, 1990, vol 2, pp 11–33.
69 Phillips RE, Rowland-Jones S, Nixon DF, et al: Human immunodeficiency virus genetic variation that can escape cytotoxic T cell recognition. Nature 1991;354:453–459.
70 Pircher H, Moskphidis A, Rohrer U, Burki K, Hengartner H, Zinkernagel RM: Viral escape by selection of cytotoxic T cell-resistant variants in vivo. Nature 1990;346: 629–633.
71 Aebischer T, Moskophidis D, Rohrer UH, Zinkernagel RM, Hengartner H: In vitro selection of lymphocytic chorio-meningitis virus escape mutants by cytotoxic T lymphocytes. Proc Natl Acad Sci USA 1991;88:11047–11051.

72 Cerundolo V, Elliott T, Bastin J, Rammensee H-G, Townsend A: The binding affinity and dissociation rates of peptides for class I major histocompatibility complex molecules. Eur J Immunol 1991;21:2069–2076.
73 Del Val M, Schlicht H-J, Ruppert T, Reddehase MJ, Koszinowski U: Efficient processing of an antigenic sequence for presentation by MHC class I molecules depends on its neighbouring residues in the protein. Cell 1991;66:1145–1153.
74 Hahn YS, Braciale V, Braciale T: Presentation of viral antigen to class I major histocompatibility complex-restricted cytotoxic T lymphocyte: Recognition of an immunodominant influenza haemagglutinin site by cytotoxic T lymphocyte is independent of the position of the site in the haemagglutinin translation product. J Exp Med 1991;174:733–736.

Rodney E. Phillips, Molecular Immunology Group,
Institute of Molecular Medicine, John Radcliffe Hospital,
Headington, Oxford OX3 9DU (UK)

Subject Index

Acquired immunodeficiency syndrome, origin 3
ADCC activity 68
Amino acid sequence
  envelope proteins 91, 92
  epitopes 158
  variability 38
Animal models for
  HIV vaccine studies 63, 64
  T cell epitopes delineation 143
Antibodies
  genetic manipulation 120–122
  neutralizing 11, 12
Antibody
  combinatorial libraries 114–119
  libraries 122–124
Antibody-binding antigenic sites 65–67
Antibody-producing cell lines 122
Antigen processing, HLA class-I binding 154, 155
Antigen-presenting cells 128
Antigenic sin mechanism 16, 17
Antigenic variation of HIV-1 9, 10
Anti-gp41 antibodies 42
Anti-gp120 antibodies 40, 113
Anti-gp160 antibodies 41
Anti-HIV-1 antibodies
  generation, recombinant DNA 116–119
  subtyping 42, 44, 45
Anti-HIV-1 vaccines, components 45–47
Antipeptide antisera, B cell epitope mapping 41, 42

B cell epitopes on viruses 36–38, 49, 50
  mapping on gp120/gp41 38–42, 44, 45
  molecular mimicry 49
Biological phenotype 12, 13
Biological variation of HIV-1 9

CB-HIV-1 gp41 104
CD4
  binding to gp110 81, 82
  binding to gp120 47, 48, 97–99, 117, 120
  receptor 11
  T cells 127–145
  T cell epitopes 127–145
CDR3 region of CD4 receptor 13
Chain shuffling 121
Chimpanzees, source of antibodies 123
Clonal analysis of T cell epitopes 135–140
Conformational escape mechanism 16
cpIII 114, 115
Cytotoxic T cell(s)
  HIV 155–157
  immunity 150–159
Cytotoxic T lymphocyte(s)
  determinants 78, 138, 139
  genetic variation 157, 158
  immunity to viruses 152, 153
  recognition of viruses 153, 154

DB4B7 monoclonal antibodies 104
Deglycosylation, effect of gp110 binding on CD4 81–83

Delta B670 viral isolates 81
Diagnostic reagents for anti-HIV-1 antibodies 42
Dilution endpoint titers 39, 40
Discontinuous virus-neutralizing epitopes 47, 48
DPw4.2 139
DR4-Dw10 clones 136
Dw4 132
Dw14 132
Dw15 132

ELI isolate 6
Endogenous retroviruses, *see* Retroviruses
Envelope coding region 6
Envelope *(env)* gene 2
  nucleotide substitution 8
  variability 6, 7
Envelope portion of SIV 78–88
Epitopes
  discontinuous virus-neutralizing 47, 48
  on gp41 100–106
  on gp120 91–100
Exogenous retroviruses, *see* Retroviruses
Evolution rate 5

F8/5E11 monoclonal antibody 87
F105 monoclonal antibody 99, 100
Fab(s), isolation 114–117
Fd region 114, 115
Fluorescein 123

Gag gene
  genetic variation 157–159
  nucleotide substitution 8
  sequence of HIV 62, 63
Gene rescue from cell lines 122
Genetic variation in HIV 5–9, 156, 157
GLG amino acid sequence 46
Glycosylation, effect of gp110 binding on CD4 81–83
N-Glycosylation sites 36
gp41
  antigenic site mapping 34–37
  disulfide loop 42–44
  epitopes 100–105

gp110
  binding to CD4 81, 82
  sera of SIV 83, 85, 86
gp120
  antibodies 113–120
  antigenic site mapping 34–37
  discontinuous virus-neutralizing epitopes 47, 48
  epitopes 91–100
  immunity to viruses 151
  immunodominant regions 92, 93
  inhibition of binding 11
gp120-IIIB 40, 41
gp120/gp41
  amino acid sequence variability 38
  B cell epitopes 38–42, 49
gp140 of SIV 83
gp160-IIIB 40, 41
GPG amino acid sequence 46
GPGR amino acid sequence 95
Group-specific antibodies 10, 11

HIV
  glycoproteins 34–50
  infection
    acute phase 13, 14
    asymptomatic phase 14–16
    $CD4^+$ T cells 129–131
    symptomatic phase 16–18
  isolates, variation 6–8
  latency 150, 151
HIV-1
  genetic variability 5–9
  glycoproteins, epitopes 91–106
  neutralizing antibodies 67, 68, 119, 120
  proteins
    $CD4^+$ T cell epitopes 127–145
    nucleotide sequence 4, 5
    phenotypic variability 9, 10
HIV-2 3
  glycoproteins 61, 72
    antibody-binding antigenic sites 65–67
  infection, epidemiology 62
  isolates
    genomic organization 62, 63
    neutralizing antibodies 67–70

nucleotide sequence 4, 5
HLA class-I binding, antigen processing 154, 155
Hypervariable domains of SIV 80, 81

IB8env monoclonal antibody 101
Immunity to viruses 151–154
Immunodominance, glycoproteins 39–41
Infection and $CD4^{+}$ T cells 129–131

Lambda phage 112, 113
Lentivirus(es)
mutation rate 5, 6
phylogeny 3–5
Low molecular weight polypeptides 155
Lymphadenopathy-associated virus 3, 4

Mac251 viral isolates 81, 85
Macaques, models for SIV 64
Major histocompatibility cells, gene products, T cell epitopes 129, 131–135, 145
MAL isolate 6
Minus-strand recombination 6
Molecular mimicry, B cell epitopes 49
Monkeys, models for SIV 64
MoAb(s) (monoclonal antibodies)
13 104
15e 97
41-7 104
50-69 103
120-16 103
257-D 95
268-D 96
386-D 96
391/95-D 96
412-D 96
418-D 96
419-D 96
447-D 95
448-D 97, 99
453-D 96
504-D 96
559/64-D 97, 99
588-D 97, 99
694/98-D 96
1125-H 99
4117-C 96
K-14 104
KK-5 87
KK-9 87
N2-4 104
N701.9b 96
recognition of epitopes 93–106
recombinant DNA methods 112–124
neutralizing 87, 88, 119, 120
Mutagenesis for clone refinement 121
Mutations 7–9, 15, 16, 129, 158, 159
rate of fixation 8, 9

Naive antibody libraries 122
Neutralization epitopes of HIV 78–88, 151
Neutralizing antibodies 10–12
HIV-2 antibody sera 67, 68
HIV-2 glycoproteins 68–71
group-specific antibodies 10, 11
type-specific antibodies 10, 11
NG3B7 monoclonal antibodies 104
Nucleotide substitution 8

OKT4, monoclonal, effect on gp110-CD4 complex 81, 82

Panning for antibodies 115–117
pComb3 system 114–116
Peptide transporter 155
Peptides
anti-HIV-1 vaccines 45–47
diagnostic reagents 42
mapping of B cell epitopes 38–41
sequence analysis 143, 144
subtyping of anti-HIV-1 antibodies 44, 45
T cell responses 140–143
Peripheral-blood mononuclear cells 8
Phagemid 114–116
Phenotypic variability of HIV-1 9, 10
Pol gene sequence of HIV 62, 63
Principal neutralization domain 11, 45–47, 78
antibodies 12
Proteasome 155
Provirus of HIV 150

Quasispecies distribution concept 7, 8

Recombinant Fabs 119, 120
Recombination, retrovirus 5, 6
Retroelements 1, 2
Retron 2
Retroviruses
  endogenous 2, 3
  exogenous 2, 3
  origin 1–3
Reverse transcriptase 1
rgp130 86
RNA viruses 1
RP297 peptide 84

SI-1 monoclonal antibody 100
Semisynthetic antibody libraries 123
Serum neutralization, HIV isolates 9, 10
Severe combined immunodeficiency mice, source of antibodies 123
Simian immunodeficiency virus (SIV) 3, 4
  animal models 64
  antibody-binding sites 67
  genomic organization 63
  neutralizing antibodies 78–88
SI-V3 variant 17
SIV-agm 4
SIV-cpz 4
SIV-mac 4
SIV-mnd 4
SIV-sm 4
Sooty mangabeys 4
Syncytium-inducing viral isolates 9
Syncytium-inducing viruses 16, 17
Synthetic antibody libraries 123

T cell epitopes in HIV-1 proteins 127–145
Transmembrane glycoproteins 65
Tropism 12, 13
Type-specific antibodies 11

V3
  domain in HIV infection 10–20
  loop of gp120 and epitopes 93, 94
  loop of HIV-1 isolates 39, 45, 69
    components in HIV-1 vaccines 45, 46
  loop of HIV-2 isolates 69, 70
  loop of SIV isolates 81, 82, 84, 86
V4 loop of SIV isolates 81, 82
V10-9 monoclonal antibody 104, 105
Vaccines 128, 138, 139
  animal models 63, 64
  viral diversity 18–20
Variable domain sequences 118
Viral protein
  U 62
  X 62
Viral replication 5, 150, 151
Virus variants 13–20

Z3 isolate 4